Advances in Underwater Technology, Ocean Science and Offshore Engineering

Volume 33

Aspect '94

Advances in Subsea Pipeline Engineering and Technology

ADVANCES IN UNDERWATER TECHNOLOGY, OCEAN SCIENCE AND OFFSHORE ENGINEERING

CONFERENCE PLANNING COMMITTEE

Alain Thain, *Health & Safety Executive* (Chairman); Phil Brown, *JP Kenny Caledonia Ltd*; Charles Ellinas, *Advanced Mechanics & Engineering Ltd*; Brian Jones, *British Gas Exploration and Production plc*; Theo Kontou, *Shell UK Exploration & Production*; Paul Rosler, *European Marine Contractors*

Advances in Underwater Technology, Ocean Science and Offshore Engineering

Volume 33

Aspect '94

Advances in Subsea Pipeline Engineering and Technology

Papers presented at a conference
organized by the Society for Underwater Technology
and held in Aberdeen, Scotland, 12–13 May 1994

SPRINGER SCIENCE+BUSINESS MEDIA, B.V.

A C.I.P. Catalogue record for this book is available from the Library of Congress

ISBN 978-94-010-4514-8 ISBN 978-94-011-1178-2 (eBook)
DOI 10.1007/978-94-011-1178-2

Printed on acid-free paper

Society for Underwater Technology

The Society was founded in 1966 to promote the further understanding of the underwater environment. It is a multi-disciplinary body with a worldwide membership of scientists and engineers who are active or have a common interest in underwater technology, ocean science and offshore engineering.

Committees

The Society has a number of Committees to study such topics as:

Diving and Submersibles
Offshore Site Investigation and Geotechnics
Environmental Forces and Physical Oceanography
Ocean Resources
Subsea Engineering and Operations
Education and Training

Conference and Seminars

An extensive programme is organized to cater for the diverse interests and needs of the membership. An annual programme usually comprises four conferences and a much greater number of one-day seminars plus evening meetings and an occasional visit to a place of technical interest. The Society has organized over 100 seminars in London, Aberdeen and other appropriate centres during the past decade. Attendance at these events is available at significantly reduced levels of registration fees for Members or staff of Corporate Members.

Publications

Proceedings of the more recent conferences have been published in this series of *Advances in Underwater Technology, Ocean Science and Offshore Engineering.* These and other publications produced separately by the Society are available through the Society to members at a reduced cost. A careers pack 'Oceans of Opportunity' has been produced by the Society in response to the growing demand by students schools and colleges for up-to-date information.

Journal

The Society's quarterly journal *Underwater Technology* caters for the whole spectrum of the inter-disciplinary interests and professional involvement of its readership. It includes papers from authoritative international sources on such subjects as:

Diving Technology and Physiology
Civil Engineering
Submersible Design and Operation
Geology and Geophysics

Subsea Systems
Naval Architecture
Marine Biology and Pollution
Oceanography
Petroleum Exploration and Production
Environmental Data

An Editorial Board has responsibility for ensuring that a high standard of quality and presentation of papers reflects a coherent and balanced coverage of the Society's diverse subject interests; through the Editorial Board, a procedure for assessment of papers is conducted.

Endowment fund

A separate fund has been established to provide tangible incentives to students to acquire knowledge and skills in underwater technology or related aspects of ocean science and offshore engineering. Postgraduate students have been sponsored to study to MSc level and subject to the growth of the fund it is hoped to extend this activity.

Awards

An annual President's Award is presented for a major achievement in underwater technology. In addition there is a series of sponsored annual awards by some Corporate Members for the best contribution to diving operations and oceanography, and for the best technical paper in the Journal

FURTHER INFORMATION

If you would like to receive further details, please contact
Society for Underwater Technology, The Memorial Building, 76 Mark Lane,
London EC3R 7JN.
Telephone: 071-481 0750; Telex: 886481 I Mar E G; Fax: 071-481 4001.

Contents

Session 1
High Pressure/High Temperature

PRESSURE SPECIFICATION BREAK PIPELINES

DR DAVID BAYLY
Pell Frischmann Engineering Ltd
172 Tottenham Court Road
London W1P 9LF

DR W D LOTH
W D Loth & Co. Ltd
Stammerham Business Centre
Capel Road, Rusper, Horsham,
West Sussex RH12 4PZ

INTRODUCTION AND BACKGROUND

Traditionally, subsea oil and gas production systems are designed on the basis that all seabed equipment located at the remote well site and the whole length of the pipeline(s) connecting the remote development to its host platform can be exposed to the maximum well-head shut-in pressure and are therefore rated accordingly.

This approach is founded on the assumption that tree valves may fail to operate or be subject to a delayed operation and may to some extent, leak even when closed. If, therefore, the host platform is no longer able to receive the flowing production stream from the subsea source (due, for example, to either a platform ESD or process shutdown), pressure will build up throughout the whole subsea and transportation system. The rate at which the pressure builds up will be dependant on

- the continued ability of production fluid to enter the seabed/transportation system through open tree valves, delayed closure of tree valves or leakage through closed tree valves
- the compressibility of the production fluid .

If a flow or leak path continues, the system will eventually reach an equilibrium condition with all parts at the maximum system pressure ie. the well-head shut-in pressure,

In the case of a gas production system, the pressure build up in the pipeline may occur over a relatively long period as "line packing" takes place. In the case of an oil production system, on the other hand, the pressure build-up will depend upon the compressibility of the fluid but may be relatively rapid.

No attempt is thus traditionally made to engineer a "pressure specification break" into the subsea production system so as to allow those parts of the system downstream of the specification break to be designed only to withstand a lower pressure.

Volume 33: Aspect '94, 3–24.

The system pressure specification has to be reduced once the production stream has been brought onto the host platform in order to facilitate the design of the separation and process plant. This will be achieved downstream of the "riser" ESD valve (on the production manifold or on/immediately prior to the first stage separator) by means of pressure relief systems (comprising pressure safety valves and/or blowdown valves) which vent the stream to the platform flare system (and/or hazardous drain treatment system). Relief at this point, in order to provide a pressure specification reduction, is thus achieved by traditionally proven techniques and in a manner which is acceptable (by necessity) even under emergency shut-down conditions.

The elements of a typical subsea production system as described above are illustrated in Figure 1.

If it were possible to ensure that a subsea pipeline was not subject to the full wellhead shut-in pressure, capital cost savings may be possible from the resulting lower pipeline design pressure requirement. Clearly this has always been the case although for the subsea developments to date in the North Sea there may not have been a significant benefit to be gained from the provision of a pressure specification break at the remote location. Thus, for example, on systems where wellhead pressures are relatively low, operating pressure may not be the overriding consideration determining pipeline wall thickness. Likewise, if the subsea development is relatively close to the host platform or has a low production rate (as is often the case for a single well tie-back) the possible CAPEX saving on pipeline cost may not have warranted the specification break.

Potential CAPEX savings may however become of increasing significance in future proposals to develop, what have been identified as the "difficult field" prospects in the North Sea. Such prospects include:

- high pressure fields - typically having wellhead shut-in pressures of 5,000 - 10,000 psi and high gas content - where pressure will be the factor determining pipeline wall thickness
- condensate and volatile oil fields - having wellhead pressures in excess of 10,000 psi and high fluid temperatures - where again pressure containment will be the overriding factor determining the pipeline design and a significant wall thickness requirement will result
- aggressive fluid properties - ie. high CO_2 and H_2S levels, possibly combined with the above high pressures and temperatures, where the requirement for corrosion resistant steel (or high corrosion allowances) combined with the requirement for heavy pipe wall will further increase pipeline material, fabrication and installation costs.
- increased remoteness from the host platform - where the increased length of pipeline will increase its CAPEX significance

Many of the existing prospects of these types must be regarded as economically marginal developments and the possible CAPEX savings resulting from a remote pressure specification reduction may not only be significant, but may be critical to the economic development of the prospect.

PRESSURE BREAK - SYSTEM OPTIONS

Under static "equilibrium" conditions, a change in pressure in any fixed volume part of a production system can only be the result of an imbalance between the material flow into that part of the system and the material flow out of that part.

There are therefore only two fundamental methods for maintaining part of a system at a lower pressure than that of an adjacent part under "no flow" conditions:-

- maintain an effective barrier at the boundary between the two sections in order to prevent an increase in the material inventory of the lower pressure system

or

- provide a relief system in the lower pressure system to reduce (or maintain) the material inventory of the lower pressure system and thereby limit the build up of pressure in that system

(Note that the above simplification by-passes two transient effects which may occur on shutting in a line. Firstly, it is assumed that flowing fluid temperatures will always be greater than shut-in (ambient) temperatures so that the line will cool following a shut-in and hence initially reduce the pressure of its contents. It is also assumed that flow requires a positive pressure gradient to be applied (to overcome friction) and hence that the shut-in pressure of a fixed volume of fluid must be less than the maximum pressure under flowing conditions.)

If either of the above methods can be adequately developed and applied, then the pressure specification break problem is solved. The problem with the first method is that the barrier must be both highly effective and reliable and have a low enough frequency of failure to represent an acceptable level of risk. The problem with the second method is that the relief system must have an acceptably high level of reliability and be environmentally acceptable.

In Fig.1, one or more subsea wells are assumed to produce into a subsea pipeline via a production manifold header. The individual wells may be choked into the manifold (as shown) and there may be other manual valves in the flowpath (not shown). The trees may be hard piped to the manifold or may be remotely connected via a further flowline.

The only valves provided for remote closure of the production stream are assumed to be the individual tree wing, master and downhole safety valves. Other flowpaths (test, utility, kill, gas lift) are not illustrated. At the platform end a (top-of) riser ESD valve is provided (which may be in addition to or in place of a subsea pipeline isolation valve in the vicinity of the riser base).

Here, no pressure specification break is provided and the subsea system (tree flowlines and manifold header), the subsea pipeline and riser are rated to the highest pressure which they may experience which will be the maximum wellhead shut-in pressure. Wellhead shut-in pressure conditions may occur in the pipeline or riser under conditions arising from the following:-

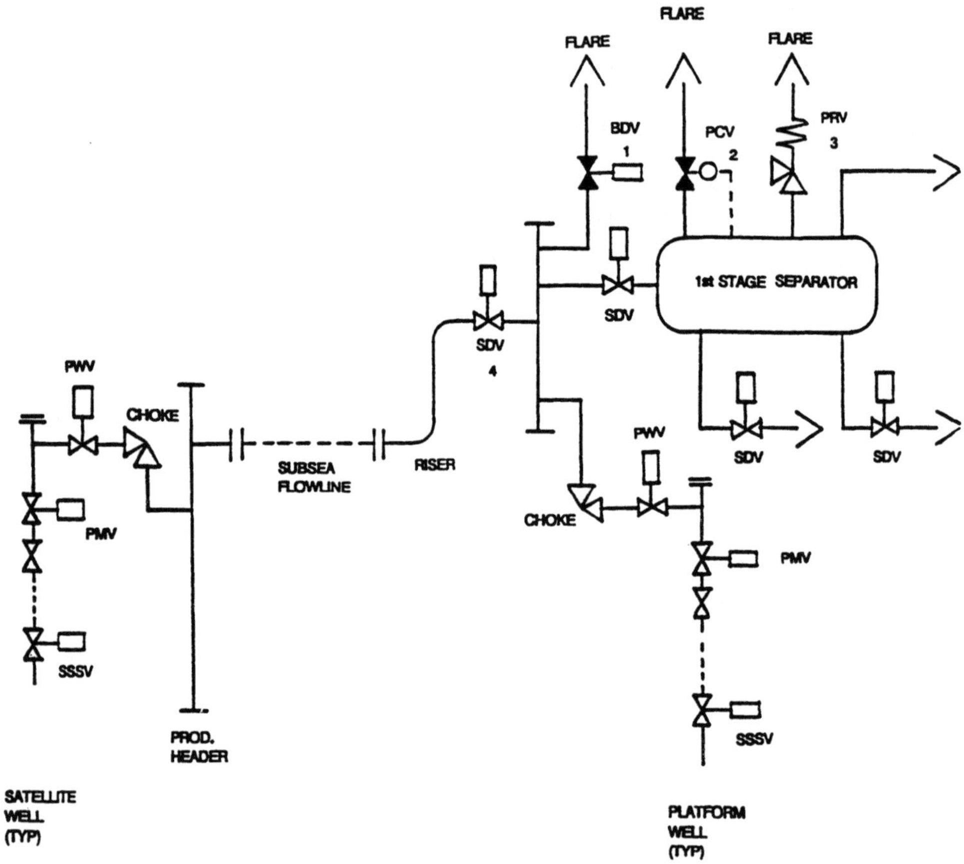

Notes: 1. Blow Down Valve (BDV) off Production Header optional
2. Pressure Control Valve (PCV) may be BDV or BDV + PCV
3. Pressure Relief Valve (PRV) may be two in parallel
4. Pipeline isolation valve may be subsea

 = Close on Platform ESD

 = Open on Platform ESD or Overpressure

Manual valves not shown (except PLM)

Figure 1 - Typical Remote Subsea System Tied Back To Host Platform

i) platform process shutdown or process condition on platform causing no (or low) acceptance of the subsea production stream whilst remote tree valves remain open

ii) closure of the platform ESD valve whilst remote tree valves (PWV + PMV + SSSV)

- remain open
- are delayed in closing
- leak when closed

iii) pipeline blockage due to

- hydrate formation
- wax build up
- liquid slug
- stuck pig

Generic methods of providing a pressure specification break are described below:-

Barrier at Upstream end of Pipeline

Tree Valves

The existing system of tree valves may be considered as the first series of barriers to prevent overpressure of a pipeline.

The wing, master and sub-surface safety valve on a tree will each be capable of remote actuation via the subsea control system using one or more subsea control modules. Whilst the subsea control system will form part of the platform ESD system, present systems will not necessarily require all (or even any) tree valves to close on all levels of platform ESD. Wing valves only may be caused to close at a particular platform shutdown level, with master and sub-surface valves only closed at the highest level (in order to eliminate unnecessary operation and so reduce wear and maintenance on these valves).

If the tree valves were relied upon to provide the pressure specification break, then the subsea control logic would require to be modified to ensure that the tree valves (wing only or wing plus master or all three valves) close on any platform ESD or process shutdown which caused the platform end of the pipeline to stop receiving production. In the case of gas or a multiphase fluid in the pipeline, a time delay between closing each end of the line would result in a limited line packing and this could be arranged not to be critical.

Closure of the remote valves only on platform command (or platform no-flow detection) would not protect the pipeline in the case of a blockage to flow occurring in the length of the pipeline or riser. To provide for protection under these circumstances, it would be necessary to provide additional instrumentation at the remote manifold to bring about the closure of the tree valves prior to the pressure building up to an unacceptable level in the manifold and pipeline. In its simplest form this would require a pressure switch at the remote end (on the manifold) to operate the tree valves (solely under its own dictates - but in addition to the ESD requirements for closing these valves). Under more sophisticated arrangements it may be desirable to close wing valves at one pressure level, master valves at a higher level and finally downhole safety valves. Careful consideration would need to be given to the necessary response time for such an instrument loop in relation to the compressibility of the production fluid and hence the pressure "rise time".

Additional Flowline Valve

As an alternative to sole reliance on the traditional arrangement of tree wing, master and downhole valves, an additional valve may be placed downstream of the wing valves (in the commingled production stream in the case of a multi-well development).

Such an arrangement has the potential advantages associated with placing a further valve in series in the production streams, but also allows the development of the additional valve to be directed solely at its pressure specification break purpose rather than placing additional functions onto the tree valves - particularly the tree wing valve.

Flowline Isolation Device

Whilst the previous section described the placing of an additional "valve" at the entry to the flowline, there is no reason to restrict this flowline isolation device to being a valve in the traditional sense.

The investment value associated with the pipeline which is being protected, together with the environmental risk associated with a possible pipeline rupture demand a consideration of all means of closing off entry to a pipeline. If this closure is used as the ultimate (rather than routine) line of defence, then even "one-trip" pipe closure devices warrant consideration.

Pressure Relief at Upstream end of Pipeline

The other fundamental method of limiting a rise in pressure in a fixed volume system is to relieve the pressure as it reaches a particular critical value by "venting" material from the fixed volume.

Relief can be applied at any point in the system, provided that there is adequate pressure communication throughout the system appropriate to the relief rate involved. This proviso also therefore requires that the relief is placed "upstream" of any position at which the system can become closed or blocked.

If blockage of the pipeline anywhere along its length (due to hydrate, wax, slug or pig) is accepted as possible, then the relief must be provided at the remote end of the system (at the remote production manifold or at the pipeline entry, ie. immediately downstream of the pressure specification break point).

Several types of pressure relieving device are available and find routine application in topside production systems (and in process systems in general). These include pressure relief valves (or safety valves), dump valves (actuated by pressure sensing instrumentation) and rupture discs. A process vessel will typically be provided with several such devices to ensure the vessel integrity.

The basic requirements of the pressure relieving system are

- that it must reliably operate at the set pressure

- that it must not operate or leak below the set pressure either prior to or following its operation (it must effectively seal and "re-seat")

- that the relieved products must be disposed of in a safe and environmentally acceptable manner (and the disposal system must be such as to not impede the operation of the relief system).

The requirement to place the pressure relief system at a remote subsea location means that requirements (i) and (ii) above will be made more critical by the difficulties of testing, inspecting and maintaining subsea components, and that requirement (iii) cannot be readily achieved by simple reliance on a platform disposal system.

Whatever the disposal system, it will be necessary to limit the amount of product to be relieved and it is therefore likely that a relief system would be provided and arranged to operate in addition to closing in the pipeline by the means discussed in the previous sub-section. Once closed in, the pressure relief system would act to limit the further pressure build up caused by continued in-flow, and the volume to be relived would only be equal to the in-flow leakage rate.

The basic options for disposal of the relieved product are

- to the sea
- to the surface
- to seabed storage
- to seabed treatment (and then dispose to sea)
- back to the host platform

and the design and acceptability of any disposal system will have to consider

- the fluid involved (oil, gas, condensate etc)
- the maximum required relief rate
- the maximum (cumulative) relief volume - possibly over a long period shut-in

To the Sea

Relief directly into the sea provides the simplest means of disposal, but raises the obvious environmental concern.

It might be considered and proven to be acceptable (non-harmful and non-hazardous) to release relatively large volumes of gas directly into the sea from a relief system. Release of liquid hydrocarbons will clearly give more cause for concern, but it may still be possible to accept such a system if the volumes are small and the probability (frequency) of release is very low.

To the Surface

Relief to the surface requires firstly a connection to the surface and secondly a surface supported system for handling and disposing of the relieved products. The disposal system would be either storage, flaring or venting depending on the fluid involved, and would require to be supported on a simple fixed or floating structure.

Any such surface structure and facility immediately raises concerns as to the integrity and station-keeping of the structure, integrity of the seabed-to-surface connection, maintenance requirements for the facility and identification/protection against shipping hazard.

To Seabed Storage

Relieved products - particularly liquid hydrocarbons - could be simply stored on the seabed at the remote location. For example, the relief stream could be piped into a bladder tank which would displace sea water as the bladder filled. This would provide containment of the relieved product without causing unacceptable back-pressure on the relief system. Facilities could also be provided to back-flow the collected products into the production system once the system was again on-stream, or it may be sufficient to retrieve and replace the storage tank.

The main problem with any such storage system will be that of determining the design storage volume - ie. the cumulative leakage volume over a long term shut-in.

To Seabed Treatment

Although seabed treatment facilities will be more complex then merely storage facilities, simple treatment may be possible which would overcome some of the problems associated with storage alone - principally the problem of necessary volume of storage.

If, for example, it was proven to be acceptable to dispose of gas but not entrained liquid hydrocarbons to the sea, then it may be possible to install a simple separation facility on the seabed to separate and release the gaseous phase and to store only the liquid phase. This would solve the problem of storing a possibly large volume of gas at sufficiently low pressure so as not to back-pressure the relief system.

It may also be possible to "treat" small volumes of liquid hydrocarbon (separated or well stream product) so as to make it acceptable to be disposed of into the sea rather than having to contain it. Even if this required a retention period such a scheme would still serve to limit storage capacity requirements.

Back to Host Platform

Transport of relieved products back to the host platform raises two problems

- the availability of a suitable sized and rated conduit to the platform other than the production line able to transport the relieved product over the distance involved without causing unacceptable back-pressure on the relief system.

- the acceptability of bringing the product onto the platform under possible platform ESD conditions.

If these problems can be overcome, then the ultimate disposal of the relived product can take advantage of the host platform vent, flare and hazardous drain systems.

Many remote subsea production systems will be provided with flowline links to the host platform other than the production line for the purpose of test, kill, utility well services, gas lift etc, and a cross connection into such a line from a pressure relief system could easily be arranged.

To be effective, the relief pipeline must be maintained at a pressure (considerably) below the set pressure of the relief system and must be available in this condition when needed by the relief system. This could not be guaranteed (or easily arranged) if both lines were in use at the time of the requirement, and a separate line dedicated to the relief function may then be justified. This could equally be a dedicated core within the control umbilical.

Also to be effective, the relief line must be able to remain open and "flowing" to the host platform under all conditions and must not therefore be subject to a common cause isolation (ESD or otherwise) together with the production line which it is designed to protect. (Some benefit would be obtained in relieving into a line shut off at the platform, but only to the extent that this line was able to "line pack" up to a pressure at which the pressure relief system performance was not adversely affected).

Under current legislation any incoming flowline carrying hydrocarbons or capable of carrying hydrocarbons has to be provided with a "riser" isolation valve (or subsea isolation valve) as part of the ESD isolation system. This is clearly a requirement on, for example, a test/kill line or gas lift line, but would strictly apply also to a dedicated relief line or an umbilical core which was intended to carry hydrocarbons onto the platform. Any such line would in effect provide a limited flow by-pass around the production line ESD valve.

The ability to relieve back to the host platform therefore relies upon the acceptability of a supported case being made to maintain an open relief line onto the platform. Some support for this could no doubt be drawn from the comparison with the relief systems provided on, for example, the platform production header serving the platform wells. Such a relief system has to remain able to relieve the accumulated leakage through platform well tree valves even under platform ESD conditions.

Pressure Relief at Downstream End of Pipeline

If the possibility of blockage along the length of the production pipeline could be totally ruled out, then it would no longer matter at which end of the pipeline the pressure relief system was located, and it would invariably be easier to locate it at the platform end.

If, for example, the pipeline ESD isolation valve was at the top of the riser on the platform (rather than on the seabed) then the relief system could also be located on the platform immediately upstream of this ESD valve.

This would clearly facilitate maintenance, inspection and testing of the relief system and provide an accessible route for disposal of the relief products.

This again highlights the fact that the ESD valve is effectively provided with a pressure relief by-pass, and again calls into question the acceptability of relief from a large capacity import pipeline as compared with that from a low capacity on-platform header or vessel.

REGULATIONS

The major legislation governing the design, construction and operation of submarine pipelines is the UK Continental Shelf area is the Petroleum and Submarine Pipelines Act 1975, Part III of which provides for a system of authorization for the construction and use of submarine pipelines and is the main control mechanism for submarine pipelines in the UK Continental Shelf Area. Regulation made under the Petroleum and Submarine Pipe-lines Act 1975, referred to as the 1975 Act, include a number of statutory instruments relating to the inspection of the lines and the limits to the jurisdiction of the 1975 Act. In respect to the design and operation of submarine pipelines, the most important statutory instrument is the Submarine Pipelines Safety Regulations 1982 SI 1982 No 1513 as amended by the Submarine Pipelines Safety (Amendment) Regulations 1986, SI 1986 No 1985. This document provides for the safe construction and operation of pipelines and for their maintenance and inspection at regular intervals. It is to be noted that SI 1982/1513 applies to submarine pipelines regardless of their date of construction.

Also appropriate to submarine pipelines are some of the Regulations made under the platform act, that is, the Mineral Workings (Offshore Installations) Act 1971. These include the Offshore Installations (Included Apparatus or Works) order 1989, SI 1989 No. 978 which has the effect of prescribing apparatus or works associated with a submarine pipeline as being within the definition of an "installation" when situated wholly or partly within 500m of an offshore installation. The other Regulation applicable to submarine pipelines is the Offshore Installations (Emergency Pipeline Valve) Regulations 1989 SI 1989 No 1029 which provide for the protection of offshore installations from the uncontrolled release of hydrocarbons from submarine pipelines by the incorporation of emergency shutdown (ESD) valves at prescribed locations in the risers.

It should be particularly noted, in relation to the pressure specification break study, that any pressure relieving or retaining systems included in a submarine pipeline system within 500m of a platform needs approval by both the Certifying Authority and the Health and Safety Executive.

The main governing regulation in respect to the pressure specification of the pipeline is set out in Clause 6 (3) of the Submarine Pipelines Safety Regulations 1982 SI 1982 No 1513. This states:-

After 19 March 1983, no controlled pipeline shall be at any time be operated unless:-

- there has been furnished by the Secretary of State for Energy a statement of maximum allowable operating pressure for the pipeline calculated by the owner of the pipeline using a formula which is based upon recognised principles and which takes account of any circumstances that will effect the operation of the pipeline at any time.

- all reasonably practicable steps have been taken to ensure that the pipeline is not operated when anything conveyed by the pipeline exerts at any point a pressure exceeding -

    (i) except under transient pressure conditions, the maximum allowable operating pressure for the pipeline.

    (ii) under such conditions, such pressure as exceeds the maximum allowable operating pressure for the pipeline by 10%.

Thus three main design constraints become apparent:-

- The maximum operating pressure for the pipeline must be determined using methods "based upon recognised principles"

- "Reasonable practicable steps" need to be taken to protect the submarine pipeline from pressures in excess of the maximum allowable operating pressure

- Pressure transients may exceed the maximum allowable operating pressure by a maximum of 10%

In order to provide guidance as to what these terms mean in practice, the Department of Energy has, since the early 1980's, published guidance notes to assist pipeline operators and pipelines designers. This document has provided background on the relevant legislation in force at any particular time together with assistance in the definition of "recognised principles" by reference to well known industry standards. The guidance notes issued up to 1989 state that "in general a submarine pipeline should be designed and constructed in such a manner that it conforms to the requirements of the "Institute of Petroleum Model Code of Safe Practice Part 6 - Petroleum Pipelines Safety Code" referred to as IP6".

Although now superseded by BS 8010, IP6 gave guidance on the use of "high integrity" protective systems for use in overpressure protection where conventional systems were unsuitable for ultimate plant protection.

Thus, to summarise, current legislation and general design guidance indicate that a pressure specification break in a subsea production system piping would be acceptable if "reasonably practical steps" were taken to protect the pipeline from pressures in excess of the maximum allowable operating pressure. Moreover IP6 suggests that barrier systems having a failure rate 0.1 times that applicable to conventional relief values would be regarded as a "reasonably practical step".

DESIGN PHILOSOPHY

A three tiered approach to overpressure protection has been considered to be the most appropriate for pressure specification break pipelines.

The first tier would be high quality, and perhaps slightly modified conventional subsea tree valves. The second tier would be a pressure and flow shut-off device activated in parallel with the tree valves but having a high level of reliability. The third tier of the safety system would be a "last chance" mechanism to prevent catastrophic failure of the pipeline. This device, which might involve direct venting to the sea, would have a very small probability of utilization. This catastrophe prevention mode might have a target probability of use once in several hundred years.

In order to allow this design philosophy to be "tested" against acceptability standards a series of bench mark analyses were undertaken to estimate over-pressure and pollution risk levels. The initial step in this process was to estimate the frequency and severity of leakage through the tree valves as currently in operation.

QUANTIFICATION OF LEAK RATES

A number of North Sea Operators were approached to see whether data was available giving leak rates measured on subsea Xmas tree valves.

The only useful data for estimating leak rates that could be found was that many operators test their valves to some variation of API 14B. There was some opinion that the maximum leak rate for oil, 400 cc/min, would be a generally acceptable criteria for valves. The equivalent leakage rate for gas was 900 SCFH.

Leakage is generally determined by bleeding down a flowline or manifold to the lowest practical pressure and using whatever well pressure is available. This leaves substantial uncertainty over the condition under which maximum allowable leakages have been observed.

In the absence of definitive data, it was decided that the leakage rates detailed above would be adopted as "typical" for "normal" pressure wells. Calculations were then undertaken, to estimate equivalent high pressure leakage rates assuming that the leakage path was of similar dimensions for both "normal" and "high" pressure valves.

Oil Leakage Rates

In order to examine the range of damage (equivalent leak path) which might lead to leakage, as specified by API 14B, flow of an API 45 degree oil was checked at pressure differentials from 1000 to 5000 psi. This led to an estimate of an equivalent orifice size at each pressure differential. The damage required to produce a 400 cc/min leak, varies from a 0.007" to 0.011" scratch. This appears reasonable for a used valve. Flows across damage of this magnitude, for a range of pressure drops, from 1000 to 6075 psi, (to match assumed pipeline pressures) were calculated to assess the minor leak rate as 179 cc/min, and the major leak rate as 988 cc/min. The logic of this approach was that damage might be expected to be similar at higher pressures as for current experience. In reality, damage might be somewhat greater although what we are concerned with is the scratching of valve internals rather than increased erosion damage; while absolute pressures may be elevated, the increased pressure drop incurred by a choke does not feature in these leakage calculations.

Gas Leakage Rates

A corresponding exercise was conducted for high pressure gas flows. Although conceptually similar, the actual calculations are more complex as the flows considered are supercritical. The fluid was the same composition as used in pipeline calculations. Using a suitably modified approach, the equivalent orifice sizes for the gas flow were calculated. Once again, the target flowrate was taken from API 14B to be 900 SCFH. The projected damage is of a similar size (to that for oil) ranging from 0.009" to 0.021". The next step was to calculate the major, 7.156 MSCFH, and minor, 1.48 MSCFH, gas flows at the higher flowing wellhead pressure of 12999 psi.

Failure Rates

Failure rate data was equally hard to locate. In the end, the most sensible course seemed to be to utilize OREDA data and convert this to probabilities for different periods. The OREDA data was supplemented with some information provided by Operators although much of this seems to stem from the same original sources.

Calculation of Leak Probabilities

The analysis was based on the assumption that all failures are random. This means that failures are independent of age or service history and implies a constant failure rate for a continuously operating system or demand failure probability independent of the number of demands. The assumption may be imperfect but not necessarily optimistic if failed valves are replaced as they normally are.

Reliability was considered for 5 and 20 year periods for oil, 5 and 10 years for gas. The effect of time in the exponential distribution stemming from the constant failure rate assumption, is illustrated in Figure 2 for a four valve system. The essential point is that the chance of failure may be greater over 20 years but the chance of a surviving valve remaining good is the same in the last ten minutes of the 20 years as the first ten minutes; that is the essential feature of the constant failure rate.

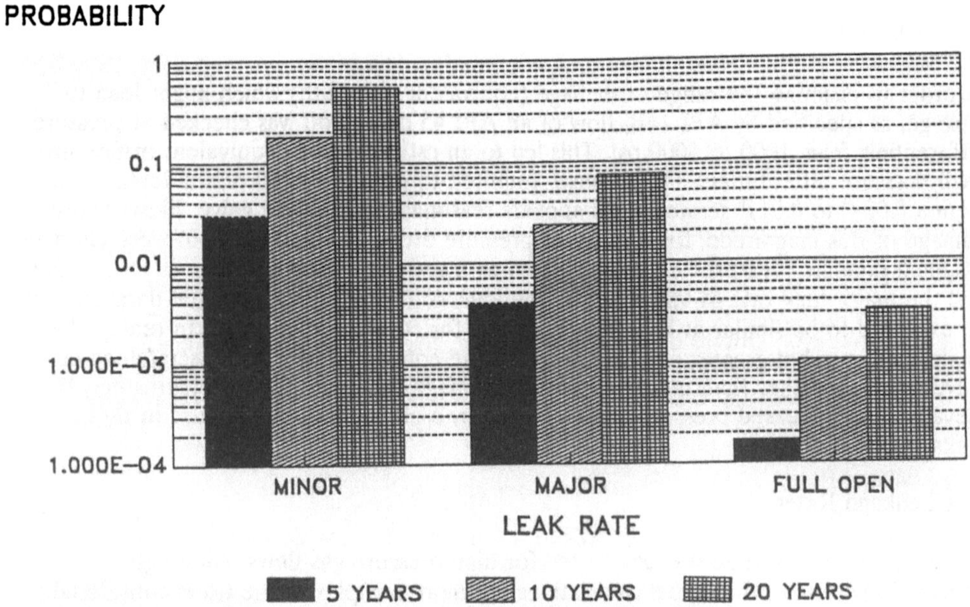

Figure 2 - Effect of Time (4 Valve Oil Production)

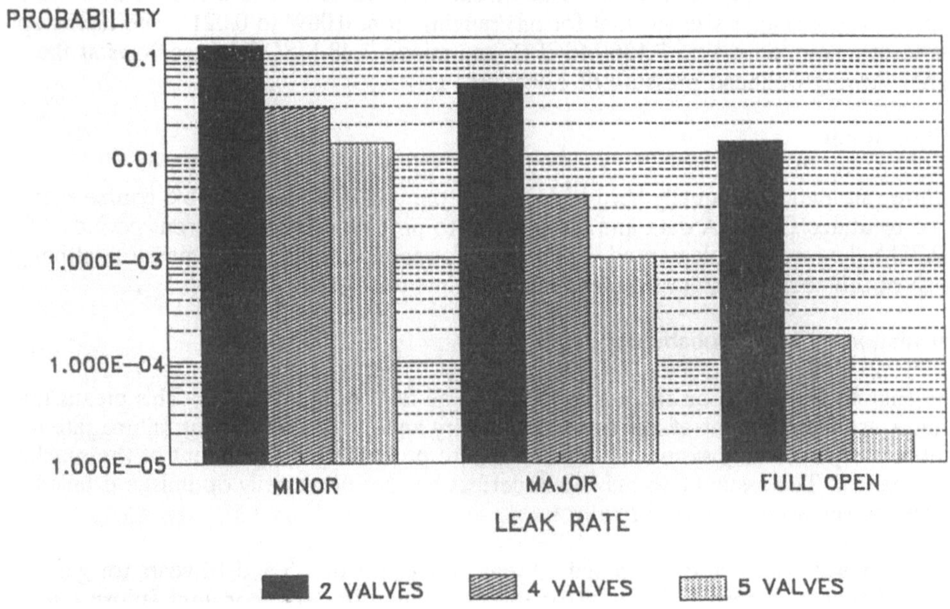

Figure 3 - Effect of Additional Valves (5 Year Oil Production)

Once individual valve failure and leak rates were projected, the next step was to consider leakage from the entire tree. Most of the work assumed that there were four valves in series. In reality there may be additional valves in series but, on the other hand, there may also be parallel leak paths through annulus and crossover valves. Leakage probabilities through 2 valve and 5 valve trees were projected to illustrate the benefit of multiple sealing possibilities. These projections are shown on Figure 3.

Each of the four valves in series could be in one of four conditions; tight, failed open, leaking at the major leak rate, or leaking at the minor rate. A single tight valve precludes any leakage into the line. Fortunately, in the majority of possible valve combinations, one or more valves will be tight. There is only the one combination, when all four valves fail open, that the well flows at its full rate. Any flowing combination which has a single valve with a minor leak rate is constrained to leak at this lower rate. In the absence of a tight valve or one with a minor leak, a combination of open and leaking valves will leak at the major leak rate.

Leakage probability has been calculated by generating all possible valve status combinations and, using the above logic, listing the leakages. This generated many probability/flowrate combinations a large portion of which have similar flowrates.

These values were used on the basis for evaluating the pressure build-up in an indicative range of pipeline applications.

EFFECT ON PIPELINES

In order to examine the possibilities of using a pressure specification break in a subsea production system a number of "benchmark" pipeline designs have been undertaken. The object of undertaking these scheme designs was:-

- To compare alternative pipeline designs based upon the pressure specification break concept with designs undertaken using the full wellhead shut-in pressure as the maximum operating pressure.
- To compare the cost of pipelines designed as in (a) above.
- To evaluate the sensitivity of typical pipelines to the estimated leakage rates and thereby estimate the probability of overpressuring or rupturing the line.

To enable the design studies to be as representative as possible two typical fluid compositions were used together with typical reservoir temperatures and pressures.

The two hydrocarbon fluids that were considered were:-

- a high pressure gas condensate
- a high pressure volatile oil (stock tank oil gravity API 45 degree)

These fluids and conditions were chosen as they exhibit high wellhead temperatures and pressures which result in the most onerous pressure conditions for the system flowlines.

System Configuration

The general case considered involved a five well development producing through a single pipeline. The wells were taken to produce at the following rates.

- Gas condensate 20 MMscf/day (stock tank conditions)
- Volatile Oil 5000 STBO/day.

For the five well system pipelines of 15, 25 and 40km were investigated.

Using the appropriate API 5L Line Pipe dimensions, the following diameters and wall thicknesses (inches) are required for flowlines with an inlet flowing pressure choked to 2880 psi, and a MAOP of 6000 psi.

Table 1 - Flowline Sizes for MAOP of 6000psi

	15 km		25 km		40 km	
	ID	WT	ID	WT	ID	WT
Gas Condensate	7.375	0.625	9.126	0.812	9.126	0.812
Volatile Oil	7.375	0.625	7.375	0.625	7.375	0.625

The API 5L line pipe dimensions (in inches) for the same length flowlines designed using the conventional method and rated for the full closed in wellhead pressure of 12999 psi (gas condensate) and 8956 psi (volatile oil) are given in the table 2 below:-

Table 2 - Pipeline Sizes rated for full wellhead pressure

	15 Km		**25 Km**		**40 Km**	
	ID	**WT**	**ID**	**WT**	**ID**	**WT**
Gas Condensate	8.625	1.298*	8.625	1.298*	10.75	1.617*
Volatile Oil	8.625	1.000	8.625	1.000	8.625	1.000

* Indicates non-standard API sizes

Pressure Build Up for Pipeline connected to Five Wells

The pressure build up times to reach RV opening, MAOP and rupture for some of the leakage rate/pipeline length combinations are given in the tables 3 and 4 below.

Table 3 - Gas

Pipeline length km	Leakage Rate (MSCF/H)	Time to RV opening (days)	Time to MAOP (days)	Time to Rupture (days)	Probability of Leak (to nearest order of magnitude) (5 year)
15	4.39	22.5	27.5	40.7	1×10^{-1}
	27.73	3.55	4.35	6.4	1×10^{-6}
	921.03	2.6 hrs	3.2 hrs	4.7 hrs	1×10^{-3}
40	4.39	84.5	103	152	1×10^{-1}
	27.73	13.38	16.4	24.3	1×10^{-6}
	921.03	9.7 hrs	11.8 hrs	17.5 hrs	1×10^{-3}

Table 4 - Volatile Oil

Pipeline length km	Leakage Rate (CC/min)	Time to RV opening (days)	Time to MAOP (days)	Time to Rupture (days)	Probability of Leak (to nearest order of magnitude) (5 year)
15	179	115	144	213	1×10^{-1}
	3502	5.9	7.4	11.0	1×10^{-6}
	553583	0.9 hrs	1.2 hrs	1.8 hrs	1×10^{-3}
40	179	562	653	966	1×10^{-1}
	3502	28.7	33.4	49	1×10^{-6}
	553583	4.4 hrs	5.1 hrs	7.5 hrs	1×10^{-3}

From the above data it can be seen that even with the considerable leakage rates that have low probability of occurrence it is a matter of days before pressure builds up to a level to overstress the pipeline. The main cause of concern is the effect of producing one well directly into the flowline in which event the pipeline will reach MAOP in approximately 54 minutes in the worst case considered.

Based upon this reasoning an estimate of the probability of pipeline overpressure and ultimately rupture has been made. Assumptions needs to be made as to the probability that pressure in the pipeline cannot be relieved at the platform end of the pipeline within certain periods of time. As indicative values to be used in estimating risk a probability of 1×10^{-1} has been assumed that a pressure build up could not be relieved within two weeks and 1×10^{-3} within two months. These are recognised as being pure estimates but are felt to be realistic bearing in mind normal operational procedures.

Based upon these figures it was determined that the probability of overstressing and subsequently rupturing the pipeline was in the order of 1×10^{-4} to 1×10^{-7} per year.

This probability may be compared with the historical data relating to loss of containment of subsea hydrocarbon pipelines. From operational experience involving 230 incidents up to 1989 (103910km-years) the probability of loss of containment of a steel pipeline is between 1 x 10^{-3} and 1 x 10^{-4} per km year. Based upon an average figure of 5 x 10^{-4} and a forty km flowline the probability of loss of containment = 2 x 10^{-2}/year.

On the basis of these figures the probability of loss of containment due solely to overpressurising of the flowline is between one and two orders of magnitude less than that arising from factors already regarded as normal in pipeline design.

It must be appreciated that all the figures above are based upon pipelines of a considerable length (15km to 40km) and blockage in the pipeline by hydrates close to the subsea system would give a higher probability of rupture. The philosophy to be adopted in thus suggested to be to use the normal valves to handle normal shut downs and use the increased safety given by a barrier valve as a guard against problems such as hydrate blockage.

RISK LEVELS

This section contains a short qualitative assessment of the risk levels associated with a pressure specification break (PSB) pipeline as compared with a pipeline designed in the normal way. The objective of this comparison is not to conclude that one or the other is safer but rather to highlight the fact that some risks may in fact be less for a pressure specification break pipeline and that a number of aspects contribute to the total risk picture.

Stress Levels

Under normal operations the stresses in a PSB pipeline will most probably be lower than those occurring in the pipeline designed for full wellhead pressure. This is because the very high wellhead pressure of the reservoirs under consideration do not give the opportunity to provide a pressure "buffer" by conservatively sizing the pipeline. It is most probable, because of the technical difficulty and cost associated with designing, fabricating and installing a pipeline for operating pressures in the range 8,000 to 12,000 psi, that the line would have high stress levels. For the PSB line it would be possible to size the pipe in such a way that the primary stresses were less.

Although it may be argued that any stress levels within the limits quoted in the relevant codes of practice are satisfactory, it is a fact that a more conservatively sized line, (such as the PSB line) has a higher capacity to resist accidental loads such as impacts from fishing gear, anchors etc.

Another factor to be considered is that at the large wall thicknesses necessary for a pipeline designed for full wellhead pressure, other effects become important and can affect the structural integrity of the line. Factors in this category are secondary stresses due to through thickness effects and high levels of residual stresses due to the welding of thick wall pipe.

Corrosion

Corrosion rates in a PSB pipeline will most probably be less than for a normal high pressure line due mostly to two factors. Firstly the Joule Thomson effect through the choke will cool the produced fluid and secondly the partial pressure of the CO_2 or H_2S components will be less. Both these effects will reduce corrosion rates in the PSB pipeline particularly in the section of pipeline directly downstream of the choke where the effects will be most pronounced.

Erosion

For fluids with a high compressibility the erosion rates will most probably be higher in the PSB pipeline due to the location of the main flow control choke close to the wellhead and the increased flow velocity in the pipeline. Care will need to be taken in the design of the PSB pipeline to ensure that a conservative design is adopted in respect to this matter.

Stored Energy

The stored energy in the high pressure pipeline would be vastly greater than that of the PSB pipeline thus the danger to personnel in the area in the event of a leak would be greater. In addition the hydrocarbon inventory in the line will be substantially less for a de-rated line.

Hydrates

Due to the slightly lower temperatures in the PSB pipeline the possibility of hydrate formation may be increased. Experience has however shown that most hydrate blockages in pipelines (as opposed to chokes or process plants) occur at locations where there are partial buckles, dents or dips in the pipeline. Thus the main contributor to actual hydrate blockage occurrence will most probably be the likelihood of local damage to the line. The increased velocity of flow in the PSB pipeline would however mitigate against the build-up of hydrates in the line.

The general conclusion to be drawn from the above is that although the need to provide overpressure protection on the line gives rise to some additional level of risk, some risk factors are reduced by operating the pipeline at lower pressure. The risk of operating a pipeline designed to take the full wellhead pressure for the high pressure/high temperature reservoirs under consideration is certainly not negligible.

DESIGN METHODOLOGY

The use of a specification break in a subsea production system would be a relatively unusual situation which would require a rather special design and verification process.

The following methodology is suggested.

- Preparation of initial scheme arrangement for the subsea production system including manifolding, valving, instrumentation and control arrangements based upon the best available information as detailed later in this section.

- Estimation, or determination from historical statistics, of the leak frequency and leak volume for appropriate valves, valve pressure differentials and fluid properties.

- Calculation of the probability of different leak rates through the wellhead valving arrangement configured in (a) above, based upon an anticipated operating philosophy (ie. the sequence of valve closing and opening).

- Sizing and design of the pipeline using normal design methodology based upon choking the flowing wellhead pressure to the normal operating pressure for the line. In accordance with accepted practice the pipeline will be sized based upon the anticipated flow conditions throughout the field life. In some instances a down-rated pipeline operating pressure may be used to give a greater "buffer" between normal operating conditions and maximum operating pressure.

- Estimation of pressure build up times in the pipeline, based upon the predicted leak rates, pipeline size and operating pressure.

- Identification of failure scenarios (ie overstressing) by the use of a failure modes and effects analysis and the estimation of availabilities and performance of protective measures for the pipeline including pressure relief at platform, vent at wellhead into dedicated or available service line or venting to sea or air.

- Determination by use of event trees of the probability of overstressing or rupture of the pipeline and probabilities of hydrocarbon releases to the environment. Particular attention will be given to common mode or common cause failures.

- Demonstration of the sensitivity of the results to different leak rates and probabilities and failure rates for system components.

- Review the probability of overstressing or releases to the environment in the light of acceptable risks in the oil industry.

This methodology allows the risk associated with the operation of a pressure specification break subsea system to be systematically examined and evaluated in the light of existing operational risk levels.

GENERAL CONCLUSIONS

It would appear that a satisfactory specification break safety system can most probably be based upon incremental improvements in existing equipment and revised operating procedures rather than development of entirely new equipment and systems.

It appears that a barrier can be an effective part of the safety system. The most salient feature of this barrier will be an extremely high probability of closure rather than an extraordinarily low leak rate.

If the data used is correct, the probability of a well failing to shut in is sufficiently high that a small bore vent system may not provide the required safety margin. In general, small leaks can be handled by pipeline design involving line packing and the vent system would be required to handle full flow from a failed well.

The most straight forward way to minimize leaks is to close multiple valves at the tree. This is a significant change in operating practice and may substantially impact the specification of valves and actuators.

The design of the pipeline can be undertaken in such a way as to provide a capability for the more probable leak rates in all but the shortest pipeline lengths.

Derating of the pipelines and operating to a marginally higher stress level can provide a useful amount of time during which leakage can be sustained. The amount of inflow, and duration of blockage, which can be tolerated is obviously dependent on line diameter and length; as these become sizeable, extended periods of inflow can be tolerated.

Risk levels, for pipeline rupture and environmental damage are not out of line with other offshore and land based operations. The most serious problem identified is the possibility of a common mode failure of all valves thus allowing full flow into the pipeline. The data used indicates that there is a probability of approximately 1×10^{-3}/year of this event occurring. The provision of an independent additional barrier valve reduces the probability of this leakage overpressurising or rupturing the line to approx. 1×10^{-6}/year even taking the assumption that no venting on the platform is possible. When consideration is given to the other likely rates, the probability of pipeline overstressing and rupture are estimated at approximately 1×10^{-4}/year to 1×10^{-7}/year for different leak rates.

The use of existing valving to provide most required protection, and the relatively modest cost anticipated for barrier valves, and material decreases in the pipeline wall thickness, indicate that substantial savings will be achieved by incorporation of the specification break. From indicative calculations for production systems involving production rates of approx 100MMscf day or 25000 STBO/day over distances of 40km the approximate savings in pipelines costs for a system with a specification break is in the order of £7.5m. If due to the corrosive nature of the fluid made likely by the high partial pressures it was found necessary to use high alloy steels for the pipeline the cost savings could be of the order of £30-35m (2205 Duplex).

Some revision of current operating philosophy, on sequential valve operations, will probably be required.

ACKNOWLEDGEMENTS

This paper is based upon a study, undertaken by Pell Frischmann Engineering in association with W D Loth and Co. Ltd which was funded by the Secretary of State for Trade and Industry and the following five industrial sponsors:-

BP International Company Ltd
Deminex UK Oil and Gas Ltd
Shell UK Exploration and Production
Texaco Britain Limited
Total Oil Marine Plc

The technical contributions of the industrial sponsors during the course of the study is gratefully acknowledged as is the assistance of the Offshore Supplies Office and the Petroleum Engineering Directorate of the Department of Trade and Industry and the Offshore Safety Division of the Health and Safety Executive.

Session 2
Design

BENDING OF PIPELINES TO HIGH LEVELS OF STRAIN

A C WALKER
Advanced Mechanics & Engineering Ltd
Surrey Research Park
Guildford, Surrey GU2 5RE
UK

ABSTRACT

There is currently a very considerable emphasis of cost saving in the installation and operating practices of sub-sea pipelines. The industry is now well established and extensive experience exists to guide developments in the use of materials and protection against failure. To a very large degree the design of pipelines has been governed by codes based on limiting conditions applied to stress. This has the effect of preventing a highly efficient use of expensive pipeline material. If the pipe geometries and operating condition can be based on limiting conditions of strain, allowing exceedance of yield stresses, a potentially significant saving on costs could be effected.

This paper presents a brief review of some information with regard to the use of strain based criteria and the bases of the limits which may apply to the use of strain criteria. The information relates to the changes of geometry and material properties which occur when pipelines are bent to high levels of strain. It is concluded generally that in certain operating conditions and installation techniques high levels of straining could be permitted without undermining the intrinsic safety of the pipeline.

INTRODUCTION

Traditionally structural design for static loading has been based on ensuring that the maximum stress which exists in the structure will attain a level which is limited to a proportion of the yield stress of the material. This approach ignores the presence of unquantifiable levels of residual stress which can exist in the structure even in conditions of zero loading. These stresses result from the manufacturing process of the steel plates, tubulars etc., and the fabrication procedures such as welding. More recently developments in design codes have recognised the more rational approach of defining limit states of structural behaviour and designing the structure such that the applied loads are limited to a fraction of the loads which would result in the limiting states occurring. These limiting states may involve strains being induced in the material which are in excess of those at which yielding may be considered to occur.

Economic advantages have been recognised for this approach to the specification of the design guidance [1] since in certain conditions the material can be used much more efficiently than is possible in the stress limiting approach. There is a penalty in the more rational approach. This lies essentially in the requirement that the limit states of behaviour

Volume 33: Aspect '94, 27–34.

must be carefully identified and ensuring that the design analysis addresses accurately the states of loading applied to the structure, or element, to ensure that the limit states are not attained. Moreover, the possibility of interactions between apparently independent limit states must be considered and the combination of types of loading should be addressed in the analysis. In other words, the design approaches as well as being rational must be based on a clear understanding of the limit states which can occur and the influence of the material properties on the load levels impinging on these limit states.

Pipelines have no degree of redundancy which would facilitate redistribution of internal loads. However, it has been demonstrated [2] that the primary limit state of the pipe, i.e. bursting due to the internal pressure of its contents, can be treated separately from other modes of failure, such as local buckling provided there is not a high degree of cyclic loading. Thus the pipeline wall thickness may be calculated on the basis of a limiting circumferential stress condition and it is possible, in certain circumstances, for the other forms of loading to be related independently to limiting states of behaviour in which a strain based method of assessment is used.

This paper reviews briefly the circumstances in which the strain based approach may be used to economic advantage and considers some information which can provide a basis for the design analysis.

STRAIN BASED METHODS

Intrinsically strain based methods may only be used when the lateral deformation of the pipeline is limited in extent by a rigid boundary. To varying extents current codes permit the use of strain based assessment methods but the limiting strain criteria are not consistent and the codes requirements are based on steels and operating and installation methods in practice in the 1980's. The exception is the recently published British Code BS 8010 Part 3 [3]. This apparently opens the way to a wider use of strain based methods and the economic advantages which could result. However, experience has shown that the exploitation of this opportunity is limited by the technical information available to perform strain based calculations. There are no comprehensive fully validated formulations which would allow engineers to evaluate the strength and safety of their pipelines at large strains. However, with a knowledge of the relative accuracy of the available information it is possible to show that strain based methods can be applied.

The availability of accurate information applied to steels in current practice would provide cost savings in the following circumstances

- installation by the reeling method could use larger strains which would enable reels with smaller radii to be mounted on smaller vessels, or current reels to carry more or larger diameter pipe;
- J-tube radii could be reduced which could allow J-tubes to be fitted into smaller platforms and permit j-tubes to be used in circumstances currently not possible;
- pipelines could be permitted to span larger lengths, provided they did not suffer resonance with vortex excitation forces, and thus reduce the cost of unnecessary intervention to eliminate some spans;
- the cost of seabed preparation could be reduced if pipelines were allowed to have higher strain levels by conforming to the seabed contours.

These cost reduction measures can be introduced provided the limit states of behaviour which accompany high levels of strain are well understood.

DEFORMATION LIMIT STATES

This section presents a brief review of some information with regard to the two main forms of deformation limits states, i.e. ovalisation and local buckling.

Ovalisation

Bending of a pipeline induces flattening, or ovalisation of its cross-section. The degree of allowable ovalisation is limited in two ways, namely

- requirement that the pipelines shape must permit the inspection or cleaning by means of a pig run through the pipe,
- the flattening exacerbates the propensity for local buckling either due to the bending loads or the net external pressure.

There are very few sufficiently accurate test results with which to validate formulae for predicting the level of ovalisation corresponding to bending strain. Figure 1 shows the results from two tests [4] in which the ovalisation of 6in. diameter pipe was measured very accurately at various levels of longitudinal strain. The difference between the two experimental curves was caused by one pipe have a significant degree of initial ovalisation. The dotted curve has been plotted using equation (1) which is presented in BS 8010.

$$f = 0.06\left[1 + \frac{D_o}{120t}\right]\left(\frac{D_o\,\varepsilon_b}{t}\right)^2 \qquad (1)$$

where

f	is ovalisation defined as $(D_{max}-D_{min})/(D_{max}+D_{min})$
ν	is Poisson's ratio
D_o	is outside diameter
t	is pipe wall thickness
ε_b	is bending strain

It may be seen that the degree of ovalisation is underestimated by the BS formula and for high levels of strain can be up to 25% in error. This is probably acceptable but unfortunately there is no guidance at present with regard to levels of ovality which are likely to occur in a pipeline which has been bent and then straightened, or when such a cycle has been repeated. Based on a few results of tests in which the residual ovality has been measured it is reasonable to calculate the residual ovality as 30% of the maximum value of ovalisation. The points marked are calculated using a numerical analysis based on a theory proposed by Ades [5]. This analysis gives results in better agreement with the test results than Eqn (1) but there is a need for a much wider range of accurate test results before firm conclusions may be reached.

BUCKLING

A prediction for the critical bending strain, ε_x, at which local buckling will occur, given by BS8010 as

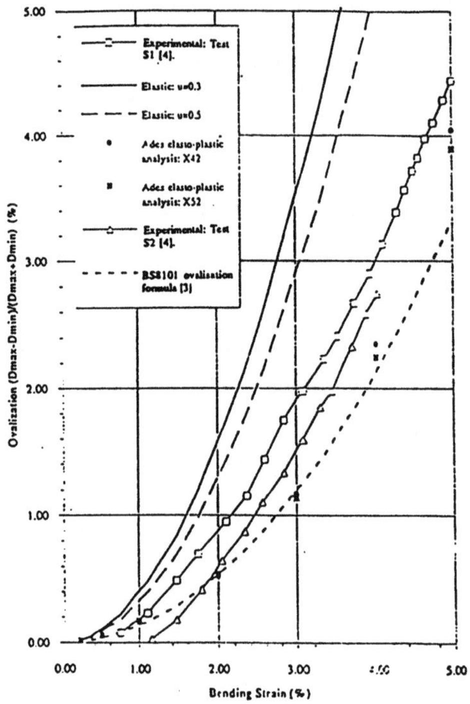

Figure 1 Ovalisation of pipes under pure bending. Comparison between experimental results of Ellinas et al [5], elastic formulae, BS8010 formula [3] and Ades type calculation [5].

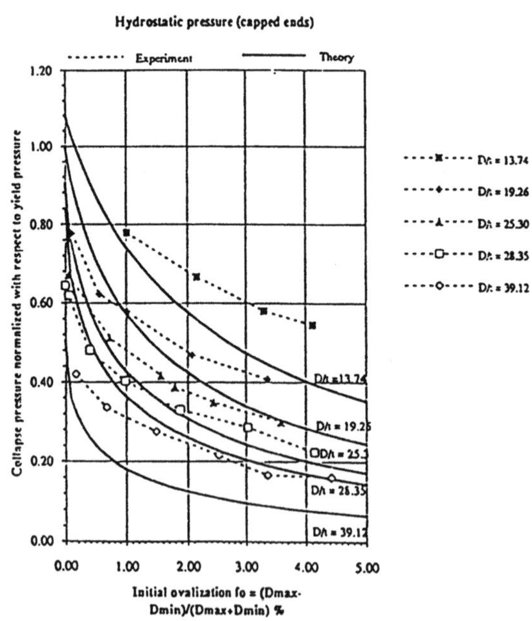

Figure 2 Pipes buckling under external pressure. Comparison between proposed generic Perry curves and experimental results for deliberately ovalised pipe from Yeh and Kyriakides [6].

$$\varepsilon_{bc} = 15\left(\frac{t}{D_o}\right)^2 \qquad (2)$$

and has been derived from the fitting of a lower bound to test results [1]. There is, however, no correlation between this critical strain and the presence of ovalisation in the pipe. This may be important especially in circumstances where the pipe may develop significant bending during operation and has been controlled by a method which impacts residual ovality at higher levels than would be expected in an as-received state.

Figure 2 shows the results of an initial analysis relating the external pressure to cause collapse to the initial ovalisation for a range of pipe geometries. The analysis is compared to tests conducted by Yeh ad Kiriakides [6] in which the pipe specimen were deliberately ovalised before being pressurised. The analysis is based on a Perry approach relating the purely plastic buckling to the corresponding elastic buckling behaviour using the plastic buckling formulae presented in reference [7].

$$(p_{cr1} - p_{cr})\,(p_{cr2} - p_{cr}) = 3f_o\left(\frac{D}{t} - 1\right)p_{cr2}\,p_{cr} \qquad (3)$$

where p = extreme pressure;
cr = collapse magnitude of p;
f_o = initial ovalisation.

The equation for moderate thick pipes is [7]

$$p_{cr1} = 2E\left(\frac{D}{t} - 1\right)\left\{\frac{\gamma}{\Delta + \frac{1}{3}\left(\frac{D}{t} - 1\right)^2}\right\} \qquad (4)$$

where

$$\gamma = \frac{\frac{4E_s}{E}}{1 + \frac{3E_s}{E_T}}$$

$$\Delta = \frac{\left\{1 + \frac{E_s}{E}(2\nu - 1)\right\}^2}{1 + \frac{3E_s}{E_T}}$$

and E = Young's modulus;
E_s = Secant modulus
E_T = tangent modulus
ν = Poisson's ratio

As regards the mode 2 buckling pressure p_{cr2} the standard elastic buckling formula is used:

$$p_{cr2} = \frac{2E}{1 - \nu^2} \left(\frac{D}{t} - 1\right)^{-3}$$

MATERIAL PROPERTIES

It is accepted that low strains, usually up to about 0.5%, will not affect the material properties significantly providing there are only a few loading cycles involving strain reversal. However, larger strains may affect the material properties, specifically

- the strength of the pipe during bending and straightening,
- the ultimate strength and ductility,
- the Charpy toughness.

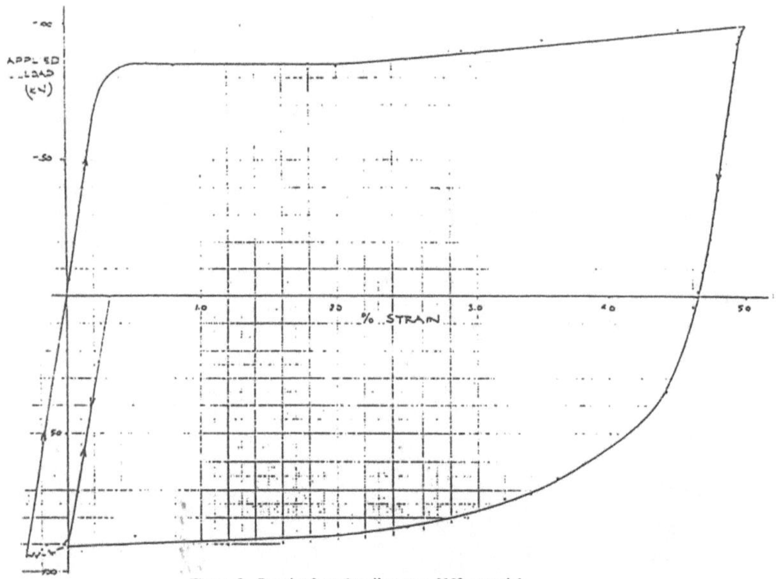

Figure 3 Results from bending test, X52 material.

This Section presents briefly the results of some testing performed on a section of pipe made from X52 material. Figure 3 shows the plot of the relationship between the applied load and measured strain on a section of pipe. The pipe was loaded in four point bending and has an outside diameter of 88.9mm and wall thickness of 7.6mm. The length subjected to constant strain was 533mm and the loading arms were 439mm long. It may be seen that the specimen had a plateau of constant moment up to a maximum strain of about 2%. Thereafter a marked strain-hardening was observed. On unloading, it may be seen that the level of the yield stress had been depressed due to the Bauschinger effect. Nevertheless, when the pipe was straightened the yield moment had increased compared to that in the as-received state.

Specimen Condition	Tensile Test Coupon Dimension (mm)	Stress at 0.5% Total Strain (N/mm^2)	Ultimate Stress (N/mm^2)	Elongation %
As received	37.9 x 15.9	366	539	57
Strained to 8% before tensile test	37.9 x 18.7	484	573	47
Strained to 8% aged for 1 hour at 200°C	38.0 x 17.9	489	574	45

Table 1 Summarised results from pipe material tests

Specimen Condition	Tensile Test Coupon Dimension (mm)	Stress at 0.5% Total Strain (N/mm^2)	Ultimate Stress (N/mm^2)	Elongation %
As received	37.8 x 18.6	377	550	38
Strained to 4% before tensile test	37.7 x 18.2	453	575	36
Strained to 4% aged for 1 hour at 200°C	37.8 x 18.1	466	575	34

Table 2 Summarised results from weld material tests

Specimen Condition	Charpy Specimen Dimensions (mm)	Charpy Energy (Joules)	Average (Joules)
As received	10 x 10	207 ; 208 ; 210	208.3
Strained to 8%	10 x 10	180 ; 206 ; 196	194.0
Strained to 8% and aged for 1 hour at 200°C	10 x 10	208 ; 194 ; 206	202.7

Table 3 Summarised results from pipe material Charpy tests (-10°C)

Specimen Condition	Charpy Specimen Dimensions (mm)	Charpy Energy (Joules)	Average (Joules)
As received	10 x 10	97 ; 154 ; 115	122.0
4% Strained	10 x 10	59 ; 64 ; 86	69.7
4% Strained aged for 1 hour at 200°C	10 x 10	7.7 ; 122 ; 69	89.3

Table 4 Summarised results from Charpy tests(-10°C) for weld material

Tables 1 and 2 show results from tensile coupons cut from a pipe made from X52 material and from a circumferential weld in the pipe. It may be seen that straining even up to 8% does not significantly affect the ultimate stress and elongation of the material. The yield stress is increased by cold working. Similar results were obtained [4] from tensile tests carried out on sections of pipe made from X42 and which had been subjected to bending strains of 5% and then straightened.

Tables 3 and 4 shows the results from Charpy impact tests on specimens prepared from X52 material and from a circumferential weld. It may be seen that the effect of straining is to reduce slightly the Charpy energy at failure but the values remain well above the Code requirements [8] which are for an average Charpy energy of 31 Joules and a corresponding minimum value of 23 Joules. These results have been confirmed by tests on Charpy specimens cut from X52 pipe which had been bent to 5% and straightened [9]

CONCLUSIONS

Currently there is an allowance in Codes that permits strains up to 2% to be imposed in controlled conditions of installation and operation of subsea pipelines. The information reviewed briefly in this paper shows that provided deformation limits are not exceeded it would be acceptable for the strain limits to be increased without endangering the safety of the pipeline or reducing its fitness-for-purpose.

REFERENCES

1 Ellinas, C P. et al. 'Limit state philosophy in pipeline design'. 3rd Offshore Mechanics and Arctic Engineering Symposium, New Orleans,1984.

2 Gresnigt, A M. 'Plastic design of buried steel pipelines in settlement areas'. Heron, 31 (4), 1986

3 BS 8010: Part 3: 1993. Code of practice for pipelines. Part 3: Pipelines Subsea : design, construction and installation.

4 Ellinas, C P. et al. 'A development in the reeling method for laying subsea pipelines'. 1st Petroleum Technology Conference, Perth, Australia, 1985.

5 Ades, C S. 'Bending strength of tubing in the plastic range'. Journ. Aeronautical Sciences, 24(8), p 605, 1957.

6 Yeh, M K. and Kyriakides, S. 'On the collapse of inelastic thick-walled tubes under external pressure'. Journ. Energy Resources Tech. Trans ASME 108930, p35, 1986.

7 Ju, G-T and Kyriakides, S. 'Bifurcation buckling versus limit load instability of elastic-plastic tubes under bending and external pressure'. Proc. IXth Int. Conf. Offshore Mechanics and Arctic Engineering, 5, p35, 1990.

8 Det norske Veritas, Rules for Submarine Pipeline, April 1981 (Reprint with corrections 1982).

9 Investigation of Structural and Material Implications of Bending Pipelines to Large Strains. Report 383.1/R/04 prepared by AME Ltd for HSE, February 1994.

A CRITICAL REVIEW OF NEAR SHORE AND LANDFALL PIPELINE DESIGN AND CONSTRUCTION

DR. T. SRISKANDARAJAH, DR. EASTER A. DAVID AND J.H.WATKINS
Brown & Root Limited
Transportation Systems Group
150 The Broadway, Wimbledon, London SW19 4DG
England

ABSTRACT

Worldwide expansion of offshore hydrocarbon exploration and the necessity to transport products to users on land presents many technical and environmental challenges. In particular, nearshore and landfall sections of submarine pipelines need to cross very difficult variable terrain, for example, shallow water, tidal flats, shipping lanes, etc. These areas are often exposed to extreme hydrodynamic conditions, geomorphical activity and other hazards. Furthermore, the increasing concern regarding ecological impact caused by the construction and operation of pipelines represents a major challenge to the industry.

This paper reviews nearshore and landfall pipeline design and construction aspects considering the above mentioned technical and ecoological problems. Recent experiences gained by Brown & Root in dealing with major pipeline landfalls throughout the world, in particular North Sea pipelines landing in the United Kingdom, mainland Europe and Norway, are incorporated in the review. This paper addresses various aspects effecting pipeline routing through nearshore and landfall areas such as shipping activity, morphological stability of the seabed, environmental and ecological considerations and governmental and local regulations or restrictions. Construction methods currently available to make landfalls through various terrain are reviewed. Advantages and disadvantages of these various methods are assessed.

1.0 INTRODUCTION

World-wide expansion of offshore hydrocarbon exploration due to the increasing demand for oil/gas necessitates a cost effective and safe transportation of hydrocarbon products to

Volume 33: Aspect '94, 35–58.

users on land. Sub-marine pipelines play an important role to this effect and although pipelines are generally accepted as the safest transportation method, many technological challenges and environmental constraints need to be addressed and overcome in design and construction to achieve safe operation for a design life of 30-50 years.

The pipeline transportation of hydrocarbon products constitutes three main segments, offshore, nearshore and landfall, and onshore. Due to technical difficulties such as crossing of shipping lanes, sand wave and mobile seabed areas, harsh hydrological conditions, etc. and global awareness on environmental aspects, the nearshore and landfall sections represent a difficult and costly construction problem. The present day economical atmosphere requires a safe cost effective, design with a minimal level of environmental impact, to be presented by the marine contractors which is able to meet with the client project schedule. As such the nearshore and landfall section of a pipeline becomes critical to the feasibility of a pipeline installation project.

In this paper a critical review of nearshore and landfall pipeline design and construction has been carried out to identify the current state-of-the-art for near shore and landfall engineering. An attempt has also been made to advocate a rational approach and a direction forward in this critical and sensitive design area for pipelines.

Following this introduction , Section 2.0 presents the outcome of the literature review that has been carried out. The requirements of present day nearshore and landfall design are discussed in Section 3.0. The identified critical aspects namely routing, construction and environment are considered in detail in Sections 4.0, 5.0 and 6.0 respectively. A systematic approach to the design and construction of nearshore and landfall pipelines is presented in Section 7.0. As a leading consultant in this area the experiences of Brown & Root Ltd. are presented in Section 8.0, along with schematic illustrations of routing, construction and environmental aspects of the near shore and landfall design. Conclusions and recommendations form Section 9.0 and the reference list is presented in Section 10.0.

2.0 LITERATURE REVIEW

The analysis of the nearshore and landfall part of a sub-marine pipeline can be divided into two different aspects. On one hand, there is the geological and hydrological environment of the nearshore and landfall area with all its associated engineering problems while on the other hand, there is the pipeline, which needs special attention with respect to its safety, both during the construction and operational phase. Literature pertaining to these two aspects have been collated and perused exhaustively (Section 10.0 lists the references of this study), and the factors that will affect the rational design methodology of nearshore and landfall sections of a sub-marine pipeline have been derived.

The following matters have drawn attention based on the literature review carried out.

- Importance of pipeline routing and surveys
- Construction method selection
- Environmental impact minimisation giving consideration to landfall site definition, geology/geomorphology, ecology etc.
- Importance of hydrological environment design parameters
- Role of technological developments
- Client/Contractor relationship and responsibilities
- Order of importance amongst the project cost and schedule, environmental impact and pipeline safety.

3.0 REQUIREMENTS OF PRESENT DAY NEAR SHORE AND LANDFALL DESIGN

The urgent and ever increasing demand for oil and gas requires fast-track design and construction of marine pipelines. At the same time public awareness on environmental matters and interest in the preservation of nature often constrains this. Moreover, the prevalent economic and political situation around the world today imposes legislative constraints and makes the clients demand more from marine contractors in the form of collective responsibility, at least during the construction phase of the pipeline. Thus, the requirements of the present day near shore and landfall design are:

- Minimise cost and schedule by cost effect design and construction
- Maximise the safety of the pipeline during construction and throughout its design life
- Minimise the environmental impact
- Investigate the use of state-of-the-art technology
- Increased responsibilities for contractors

In the following sections, the most important aspects namely routing, construction methods and environmental aspects, which when considered carefully will make present day requirements feasible, are discussed in detail.

4.0 ROUTING CONSIDERATIONS

In the design and construction of the nearshore and landfall section of a submarine pipeline, route optimisation plays an important role. In the near shore and landfall section risk to the pipelines arises from shipping activity, sand wave and unstable seabed locations, fishing activities, ammunition dumps and formerly mined areas, surf zones, severe hydrological environment parameters of the area etc. Further to this, restrictions to a direct line pipeline routing may also be imposed by the appropriate local governments and international organisations, in the form of environmentally sensitive areas and protected zones.

Thus, careful optimisation of the near shore and landfall section of a submarine pipeline, effected from the feasibility study stage through to the detailed design and operation phases, via interactive monitoring by surveys of the seabed and its soil characteristics, supplemented by

appropriate engineering analyses will significantly minimise the environmental impact, minimise the project investment costs and construction schedules and above all increase the safety of the pipeline during its design life.

The expensive and complex process of route finalisation, for the near shore and landfall section of a submarine pipeline can be accomplished through planned operations and phased studies. Based on the experience gained in this area of engineering work the following mandatory studies are advocated.

- Geology/geomorphologic assessment to identify the design life pipeline on seabed profile
- Hydrological data collection and flow modelling to study the stability of the pipeline on seabed and to calculate the cover requirements
- Environmental data collection and mapping to minimise impact and identify the hazard locations
- Risk evaluations, sensitivity analyses and mitigation measures
- Sea bottom roughness and soil data assessment to quantify the presweeping, rock dumping requirements etc.
- Cost estimation and cost sensitivity analyses
- Advance fatigue damage calculations

5.0 CONSTRUCTION CONSIDERATIONS

The choice of construction method used for a pipeline landfall is critical to the technical feasibility, cost effectiveness and environmental acceptability of a pipeline installation project. In order to achieve an optimised solution the various available methods should be assesed and their relative advantages and disadvantages compared. Landfall construction methods must be devised for the following three regions.

- Nearshore
- Shore Approach and Crossing
- Onshore

Nearshore Construction Methods

In many landfall situations offshore pipelay can not continue up to the shore approach without additional work or different construction methods being adopted to take account of shallow water conditions. The following are possible techniques which may be required to overcome associated difficulties.

- Dredging to allow barge access and to achieve adequate pipeline burial.
- Employ separate shallow water spread to perform pipelay.
- Combination of dredging, pipelay and backfill vessels to lay in tandem.
- Pipe string tow-out method.

In the nearshore region dredging is often required to provide on-bottom stability or protection from accidental damage. In many instances there is little other alternative as the burial depth required is beyond the range of conventional trenching techniques. The dredging vessels used would be either cutter suction or trailing suction, depending on the water depth prevalent. The cost of dredging usually forms a significant part of the cost of landfall construction, hence a minimisation of this activity is advantageous.

For many potential routes there is no alternative but to dredge large volumes. In such cases the cost of dredging and the problems of scheduling the work within an allowable weather window may affect route feasibility. In addition the impact of large scale dredging on established flow patterns or the impact on the environment of disturbed sediment may be unacceptable. It is therefore of great importance to optimise dredging volumes, hence the minimum predicted design life seabed profile, the amount of sediment infill and the stable side slope achievable must all be accurately defined. In the near shore area substantial dredging may be required to provide access for an offshore lay barge. An alternative to dredging in this case is to employ a separate shallow water pipelay spread. The relative benefit of reducing dredging by this method should be addressed.

In extreme cases, such as pipelay in estuaries or tidal flats, substantial dredging may be required even for a shallow draught lay barge. Inability to keep open a long dredged trench for pipelay, or the environmental impact of the same, may impact route feasibility. One solution is to perform dredging, pipelay and backfill in one continuous operation, as illustrated in Figure 5.1. This method, although greatly reducing environmental impact, must be carefully considered in terms of pipeline and vessel safety.

The pipe string tow-out installation method may be used in the near shore area beyond its normal application at the shore approach. One application for such a method would be in shallow water where a lay barge requires substantial dredging to reach the shore approach section. Burial of the pipe string could be by dredging or trenching, although if dredging is to be avoided the burial depth must be limited.

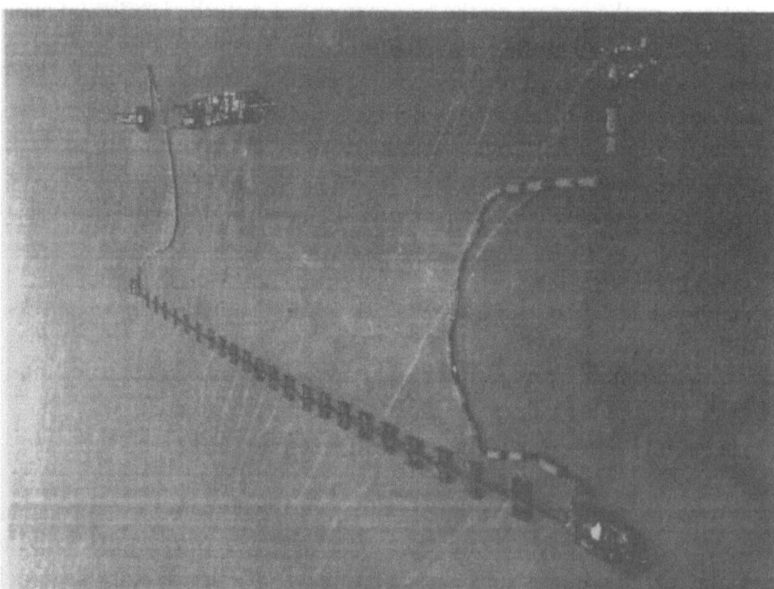

Suction Cutter Dredger, Laybarge and Backfill Spread at High Water

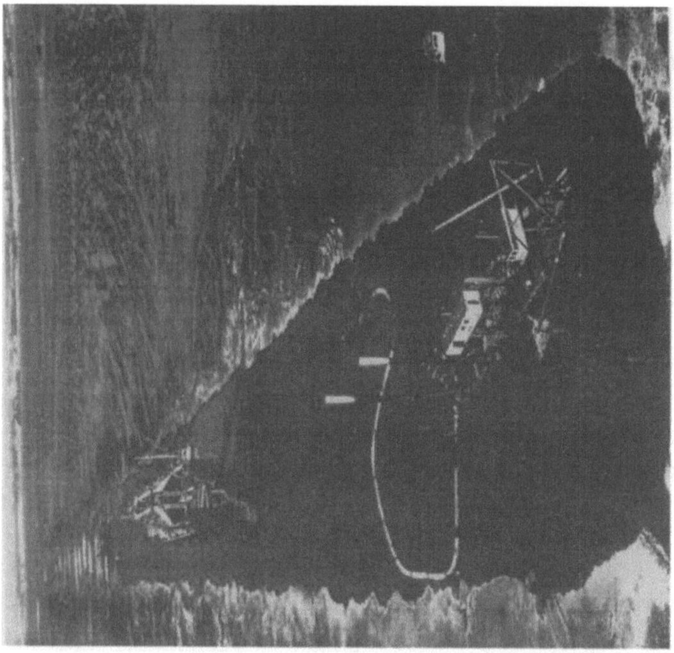

Suction Cutter Dredger, Laybarge and Backfill Spread at Low Water

Figure 5.1

Shore Crossing Construction Methods

Construction methods employed at the shore approach can be split into two broad groups, traditional open cut techniques and the more advanced closed methods. The two main open cut methods are:

- Pipe string fabricated on a lay barge pulled to shore using shore based winches.
- Pipe string fabricated onshore pulled offshore using a pull barge/pipelay barge.

Because of burial depths required at the shore both installation methods are likely to require a pull into a dredged trench. At the shore line and through the surf zone a cofferdam may be required. For a pull to shore the onshore works are restricted to a winch site of limited size, however, if the pipe is strung onshore a much larger onshore site would be required. Both these construction methods have the advantage of being inexpensive using proven technology. The main disadvantages are the impact on the environment and the potential safety implications due to routing through a high risk area. Closed construction methods overcome these disadvantages by placing the pipeline below environmentally sensitive or high risk areas. However, the limitations to the closed construction methods come from the limitations on length and associated time of construction. The two main closed construction methods are:

- Horizontal directional drilling.
- Tunnelling.

The directional drilling method is limited to approximately 1 km, while the tunnelling option can extend to around 12 km.

Both of the above methods are more technically advanced than the open cut methods and as such require considerably more engineering evaluation. In directional drilling a pilot hole is drilled from shore to a selected exit point offshore and a washover pipe inserted. A pre-fabricated pipe string, floated out to the exit hole is connected via a cutting tool, reamer and swivel to the washover pipe. The pipe string is then pulled by the washover pipe back through the hole which is widened by the cutting tool and reamer. Although expensive this method should be considered as a means to avoid both environmental impact and risk. The onshore environmental impact may be large as a pipe stringing site and drilling / pull site must be accommodated. The feasibility of horizontal directional drilling is highly dependant on the soil conditions and the length of crossing required and as such is not possible on many landfalls.

Tunnelling would be used to construct a pipeline landfall for similar reasons to directional drilling. As for directional drilling it would generally only be used where conventional directional drilling, tunnelling may be the only alternative to open cut methods.

Onshore Construction Methods

In many landfall locations the onshore section must be considered as an integral part of the landfall design. The presence of dikes, dunes or other environmental or physical obstructions may force the use of closed construction methods as detailed above, even though they are not required at the shore approach. The suitability of onshore sites with respect to environmental and safety aspects must be considered.

Landfall Construction Studies

During the selection of construction methods for a pipeline landfall the following aspects are considered of particular importance and warrant specialist studies.

- Dredging, dispersal and backfill philosophy.
- Dredging volume based on morphological and vessel constraints.
- Construction vessel selection.
- Dynamic pipelay analyses.
- Tie-in philosophy and analyses.
- Dynamic mooring analyses.
- Route profile freespan / stress analysis and intervention work definition.
- Shore approach philosophy based on routing and environmental restrictions, cost and schedule.
- Winch sizing and onshore site layout.
- Construction procedures based on safety and environmental impact studies.

6.0 ENVIRONMENTAL CONSIDERATIONS

The literature reviewed clearly showed that in the past the hydrological environment was mainly considered by pipeline designers and only recently due consideration has been given to the geological and ecological environments that encompass the near shore and landfall sections of the pipeline. To guarantee the safe operations of a pipeline in the near shore and landfall area with minimal environmental impact the following aspects need to be studied in detail.

Hydrological design parameters

Water depth measurements will enable the designer to assess the stability of the pipeline, assess different construction methods and to select the appropriate pipelay and dredging vessels with supporting equipment.

Seabed topography and sea bottom soil characteristics when acquired correctly will determine the trenchability, liquefaction potential, rate of soil erosion, scour development and sediment transport, friction factors to use and the stresses induced in the pipeline during trenching operation etc.

Wind, wave, currents, storm surge, etc. need to be gathered in a precise manner to study the installation feasibility, the on-bottom stability and the induced fatigue damage of the pipeline.

Geological Hazards

The need for a detailed geological study of the proposed pipeline route has been bought out earlier in the routing considerations section. Some of the geological hazards that will affect the pipeline integrity are: adverse sediment conditions due to poor foundation characteristics, gas charged sediments, sediments prone to liquefaction etc., potential areas for scour development, eroding shoreline, buried channels, delta regions, sand wave areas, active faults and seismic activity regions. Careful considerations to the above mentioned aspects will alleviate the risk to the pipeline and also reduce hidden project investment and maintenance costs.

Environmental and Ecological Matters

Due to increased awareness and interest in environmental matters from the public, constructing pipelines in the near shore and landfall areas becomes a complex procedure. The following points must be studied with a view to the submission of an environmental impact assessment document, which is mandatory requirement.

- Sensitive environmental areas such as spawning, nursery areas of fish, seals etc.
- Protected zones and national parks
- Breeding and resting areas for migratory birds and other endangered birds and species
- Unique and small habitats including, but not limited to, flora and fauna, coral reef, seagrass beds, mussel beds etc.
- Recreational and monumental sites
- Anchor zones, dumping sites etc.
- Urban area minimisation to reduce noise level.

Since the time period involved for obtaining permits from the regulatory bodies and appropriate local authorities is long, the relevant environmental assessment, legislative requirements and regulations and publications and gathering views from the public should be initiated at the project planning stage itself. A prudent step would be to give proper attention to post construction activities and monitoring.

7.0 SYSTEMATIC APPROACH TO NEAR SHORE AND LANDFALL DESIGN

Every near shore and landfall section of a sub-marine pipeline is unique in its own way and a standardised design and construction methodology cannot be provided. Nevertheless, a systematic approach can be adopted so that the governing parameters, i.e cost, schedule,environmental impact and safety can be optimised. The salient features of the systematic approach to near shore and landfall design are:

- Attention to all above aspects from project inception
- Optimisation through phased studies
  - Pre conceptual feasibility studies with provision to change/reject the selected option
  - Conceptual studies
  - Detailed design phase
  - Construction and operating phase.
- Priorities and objectives set at each level
- Important aspects such as routing, construction method and environmental considerations be considered at each stage until completion
- Incorporate state-of-the-art technology if advantageous
- The role of single contractor liaising for the client
  - To minimise time necessary to obtain permits
  - To develop better communication between client, regulatory bodies and public
  - To make fast-track projects feasible
  - Commitments and responsibilities of contractors increase
- Advocate risk management and mitigation procedures
- Stipulate design life monitoring system.

8.0 BROWN & ROOT'S EXPERIENCES

The aim of this section is to highlight the importance of the identified critical aspects namely, routing, construction and environmental considerations through three nearshore and landfall design and construction project experiences of Brown & Root. Schematic illustrations are included in this section to get a better understanding of the aspect illustrated.

The work carried out for Project I posed considerable technical challenge in routing the pipeline through narrow shipping estuaries, shifting sand banks, large tidal variations, shall waters and mudflats. Also, the National Park along the Wadden sea and the protected area have been given due considerations in the landfall route selection process. A typical nearshore and landfall pipeline route is schematically depicted in Figure 8.1. From the experiences gained it can be advocated that all feasible ptions need to be addressed for a

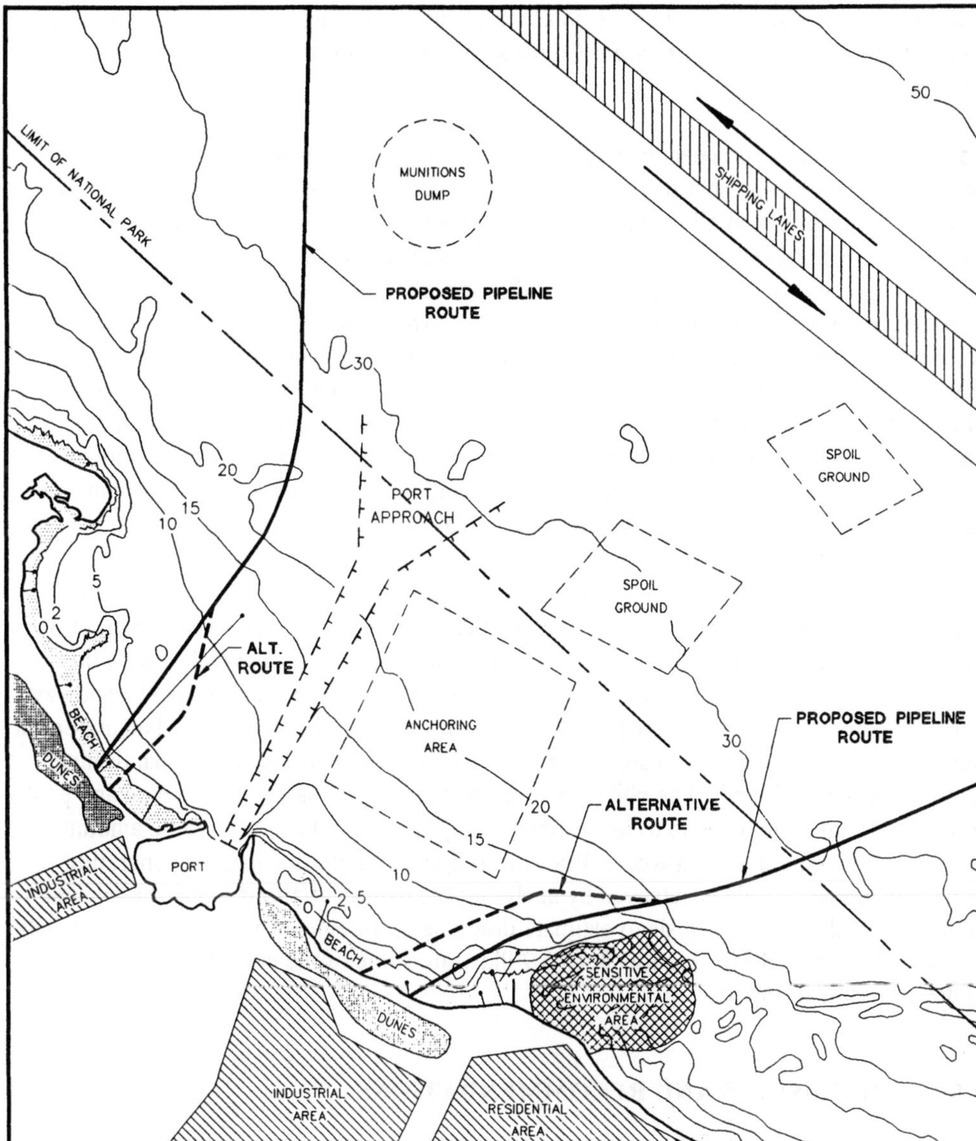

Figure 8.1 : SELECTED LANDFALL ROUTES

nearshore and landfall study, to save considerable amount of project schedule time and money, prior to applying for consent from appropriate authorities to design and operate the pipeline.

As pointed out in the preceeding sections of this paper the environmental issues are of prime importance in present day nearshore and landfall pipeline design and construction. Project II's pipeline landfall in the ecologically sensitive Coatham Sands demonstrated that the success in bring hydrocarbon reserves to this area relied mainly on the environmental methods, which were successfully applied throughout that project. An important point to be noted is that construction and restoration plans which took into account pipeline safety, engineering feasibility, and causing the least disturbance to the community and the environment, were first reviewed and accepted by local and national environmental organisations to ensure the minimum disruption to site, and reinstatement as close as possible to the original condition, prior to getting approval from Government bodies. Figure 8.2 shows clearly how a nearshore and landfall pipeline negotiates environmentally sensitive areas, before making a landfall on the mainland.

Project III in southern North Sea shows that construction activities are crucial to the design and construction of nearshore and landfall pipeline. The large scale construction operation successfully carried out included artificial island creation, winch site on the beach, sheetpiled cofferdam, barge access channel, crossing of Scheur channel, sandwave correction by dredging and trenching, rockdump rockfill requirements etc. The important point to note is that after extensive soil investigations landfall using directional drilling methods was abandoned. The philosophy of pre-dredging to safely lay the pipe and post-trenching to take the pipeline to its destined final profile saved a considerable amount of money by reducing dredging work. Prior to dredging activity the pipelines operating condition seabed profile is achieved by analysing the depth variations along the pipeline route. Seabed profiles analysis to estimate dredging requirements are show schematically in Figure 8.3A and 8.3B. Figures 8.4 - 8.7 illustrate some of the construction techniques addressed in Section 5.0.

9.0 CONCLUSIONS AND RECOMMENDATIONS

The critical review of nearshore and landfall pipeline design and construction carried out revealed that all the three aspects namely, routing, environmental, and construction should be studied properly to achieve a risk and environmental impact minimised, pipeline safety improved and a cost effective solution. Also identified was the requirement to use state-of-the-art technology and the increased responsibilities and involvement of the design and construction contractors. For a successful outcome in the design and construction of a nearshore and landfall pipeline the rational approach derived in this report is recommended.

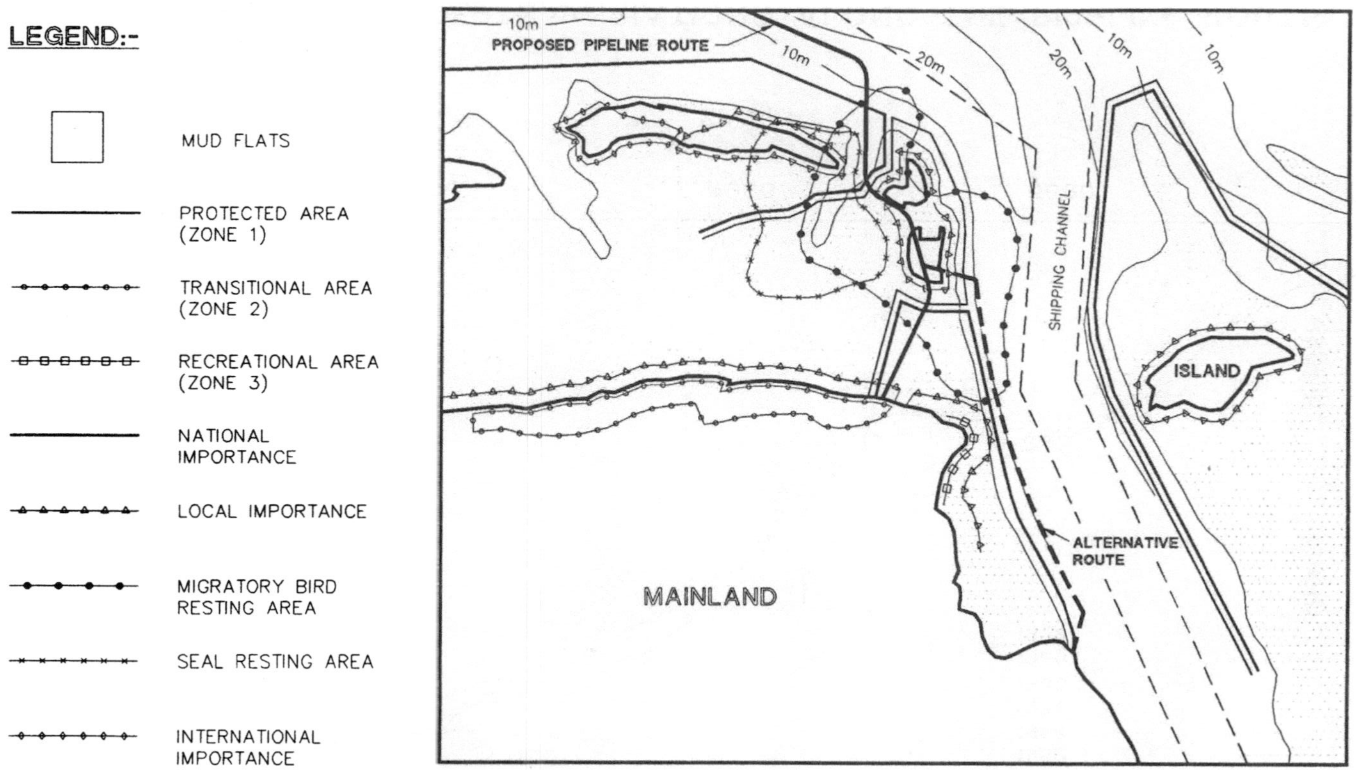

Figure 8.2 : PIPELINE ROUTE THROUGH ENVIRONMENTALLY SENSITIVE AREA

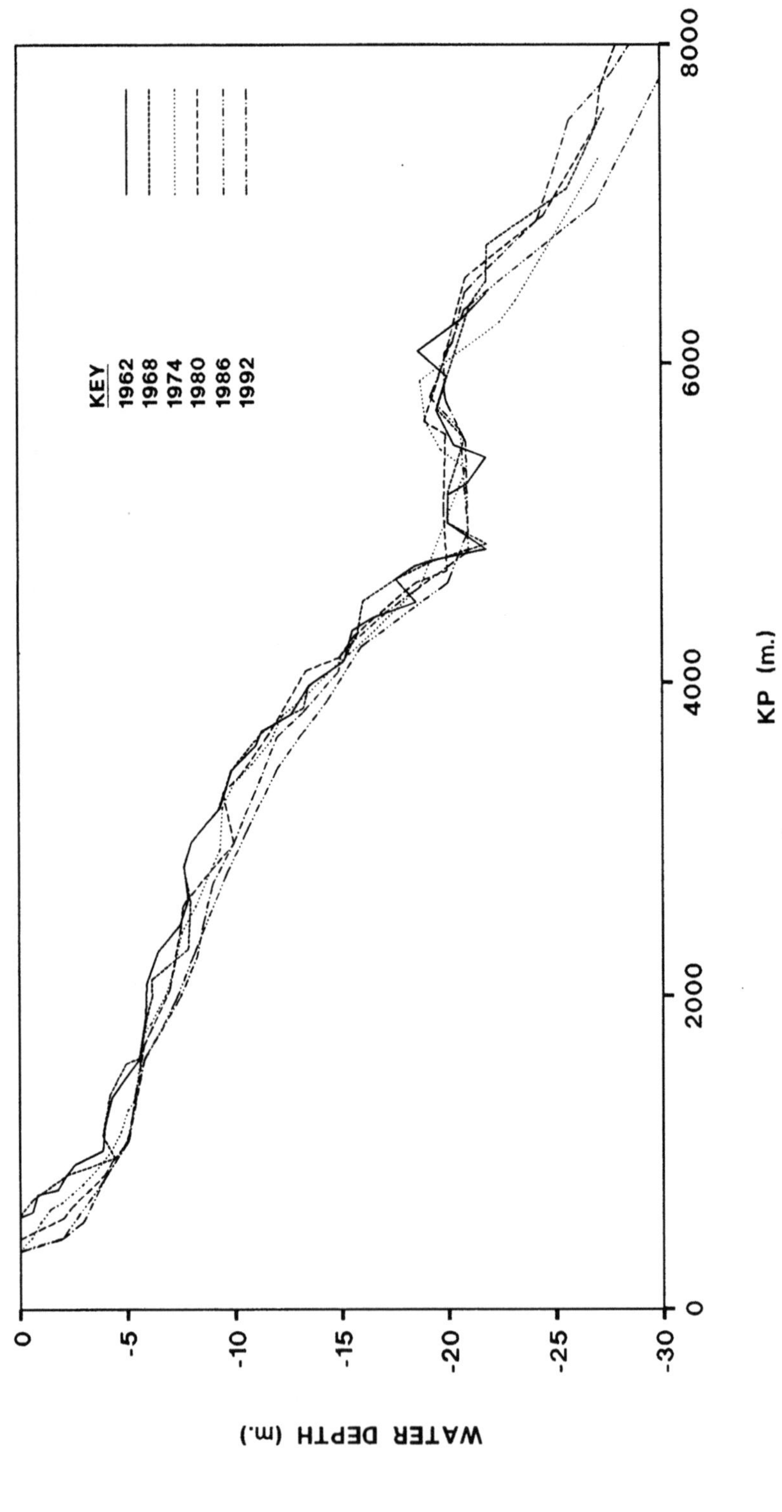

Figure 8.3A : DEPTH VARIATIONS ALONG THE PIPELINE ROUTE

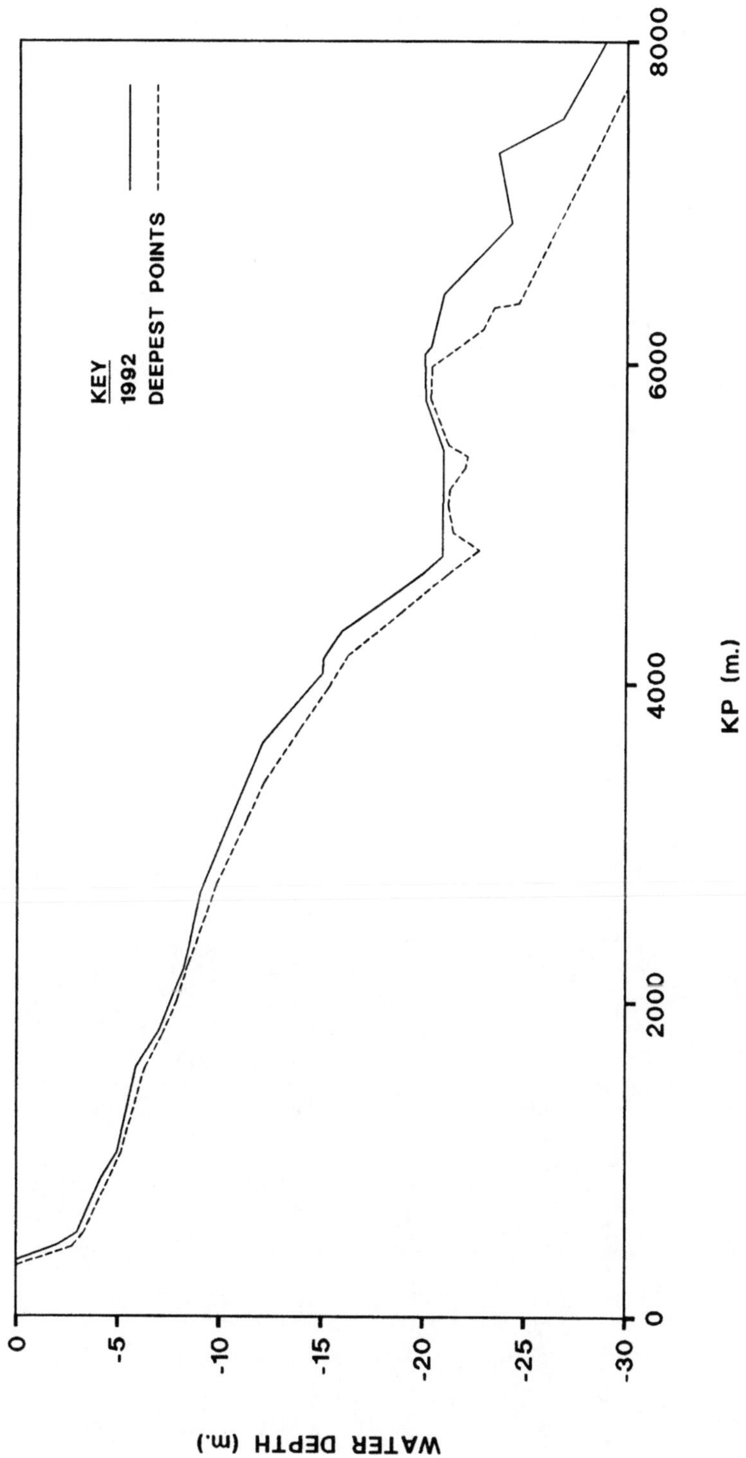

Figure 8.3B : DEEPEST POINTS SEABED PROFILE vs 1992 SEABED PROFILE

Figure 8.4 Conventional Construction Methods

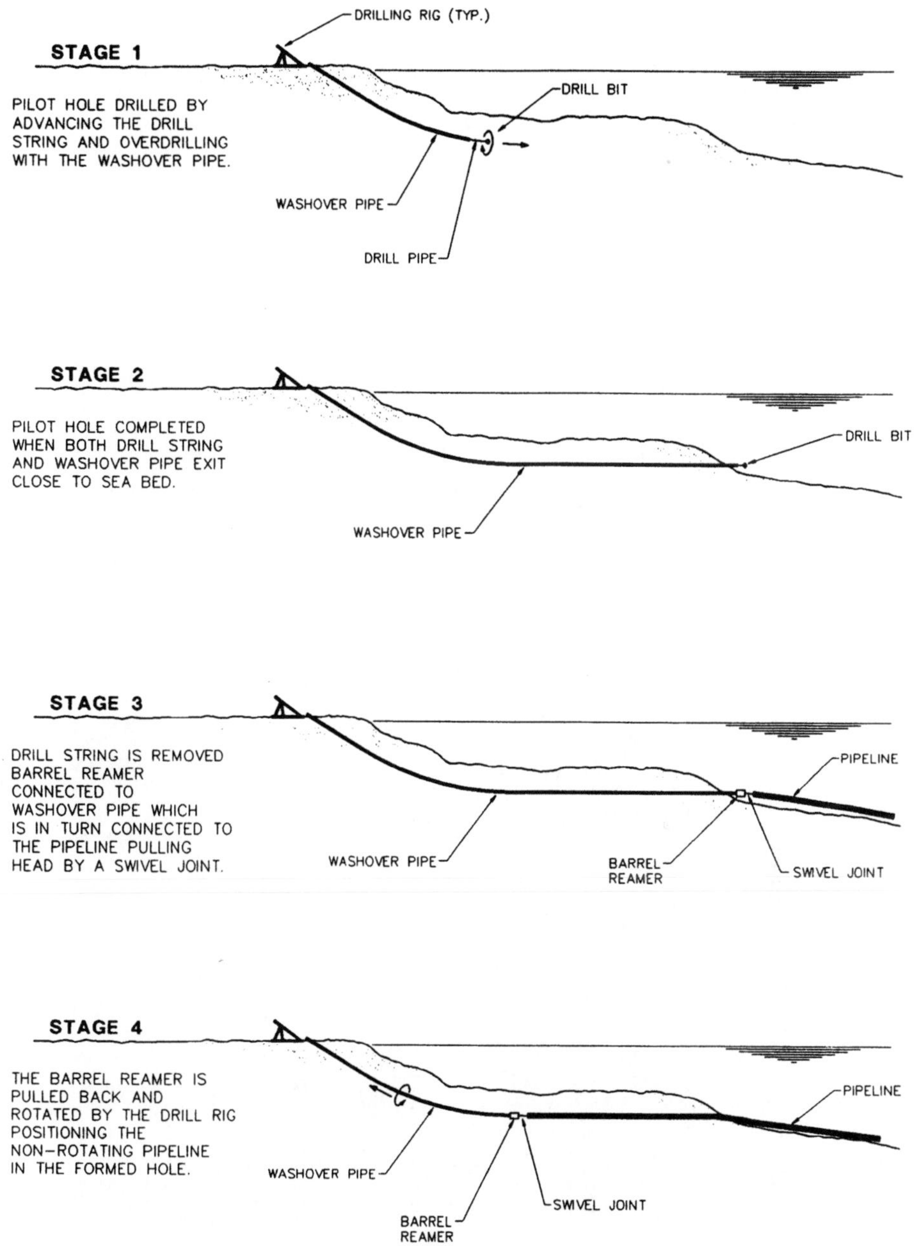

Figure 8.5A : SCHEMATIC OF DIRECTIONAL DRILLING

Figure 8.5B DRILLING PILOT HOLE FROM CLIFF TOP AREA AT BACTON

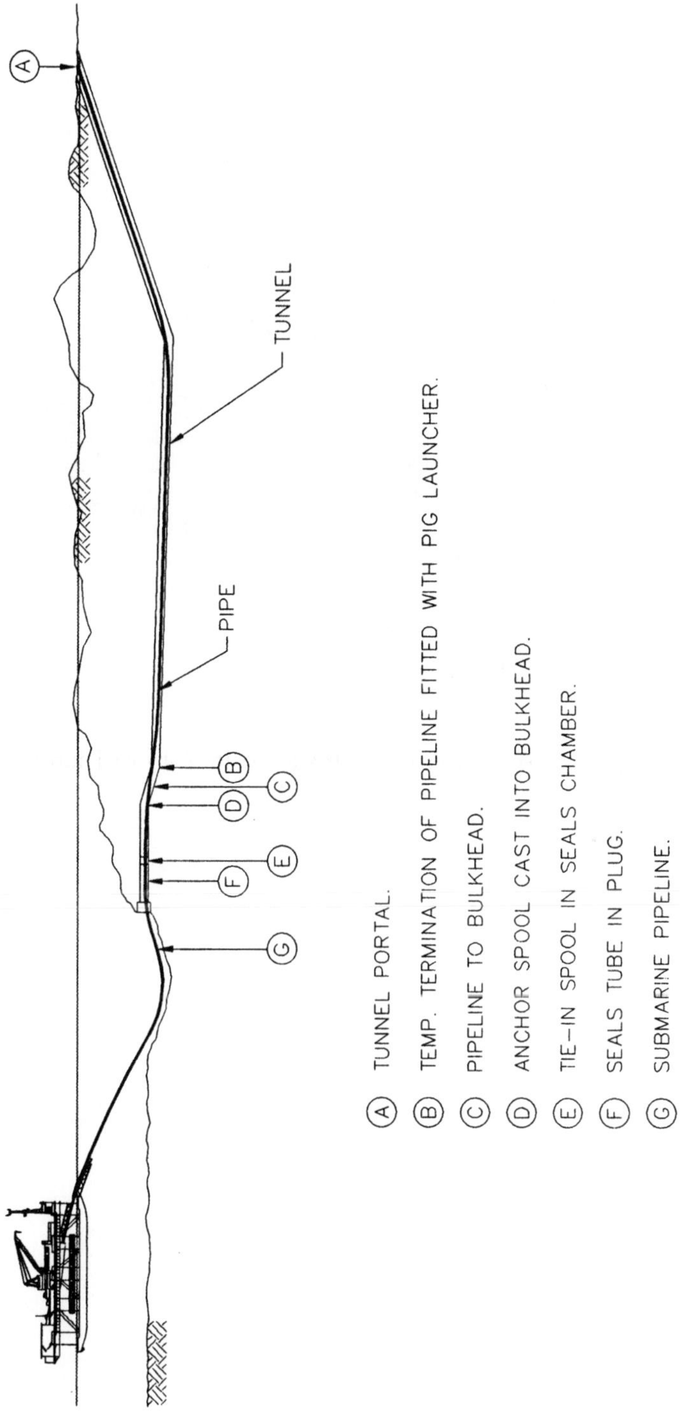

Figure 8.6A : SCHEMATIC OF TUNNELLING

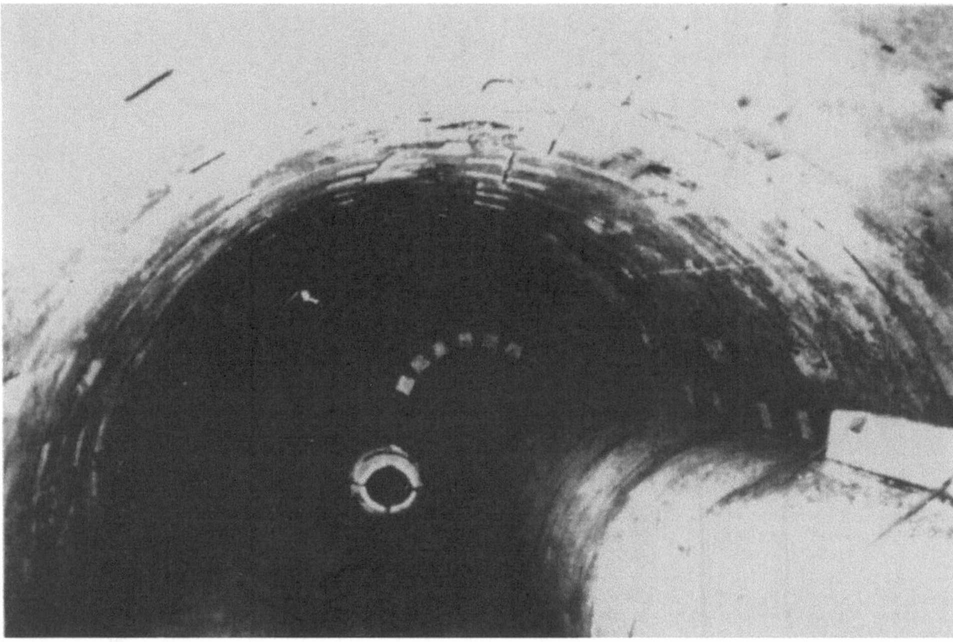

Figure 8.6B Photograph of 36" Gas Pipeline Installed in 2.44 m Diameter Tunnel

Landfall Construction at Zeebrugge

Landfall Construction at Zeebrugge

Figure 8.7

10.0 REFERENCES

d'Angremond, K. and Hiujssoon, J. A., "Pipelines in shore approaches", Pipes & Pipelines International, August 1977, pp. 19-24.

Gowen, A. W. and Goetz, N. J., "Offshore pipeline routing: minimising environmental effects", Coastal Zone '80, pp. 667-683.

"Geotechnical considerations in submarine pipeline design", Advances in Offshore Oil & Gas Pipeline Technology, 1985, pp. 1-13.

Hoel, S., "Oseberg Transportation System: 'Strandlogg Hjarety' - Shore Approach at Hjartoy", Journal Explos. Engg., 1990, 7(6), pp. 38-39.

Eide, A. et al., "Assessment of coastal processes for the design and the construction of the Zeepipe Landfall in Zeebrugge", Coastal Engineering 1992, pp. 2029-2042.

Jinsi, B. K. "Designing, constructing shore approaches for submarine lines", Oil & Gas Journal, November 1982, pp. 62-68.

Lund, S. et al., "Statpipe shore approach", Offshore Oil & Gas Pipeline Technology, January 1984, pp. 1-38.

Etter, C. K. and Bettelley, L. M., "Approval and restoration of the environmentally sensitive CATS pipeline landfall", 25th Annual Offshore Technology Conference, 1993, OTC 7170, pp. 411-420.

Larminie, F. G., "Environmental Aspects of Deep Water Large Diameter Pipelines - North Sea", UNEP Environmental Consultative Committee on Petroleum Industry, June 1981, pp. 1-13.

"McDermott makes first pipe line crossing of the Norwegian Trench", Pipeline Industry, July 1984, pp. 31-32.

Brat, J. B.,"How a Dutch crude line made landfall', Pipeline Industry, March 1986, pp. 65-68.

Lund, S. and Gjertveit, E., "Statpipe experience reveals techniques for seabed problems", Oil & Gas Journal, April 1986, pp. 57-62.

Chaney, G. J. and Smith, N. D., "U.K. pipeline installation meets environmental demands", Oil & Gas Journal, July 1989, pp. 49-56.

Schaap, P. A., "NOGAT line opens northern Dutch gas fields", Ocean Industry, April/May 1990, pp. 88-92.

Harneshang, K., "Statpipe subsea shore approach bridge tunnel - its damage design modifications and repair", Proceedings of the Third International Offshore and Polar Engineering Conference, Int. Soc. of Offshore and Polar Engineers, 1993, pp. 220-225.

Kaustinen, O. M. et al., "Submarine pipeline crossing of the M'clure strait", VTT Symposium (Valtion TeknillinenTutkimuskeskus) V27, VTT, Espoo, Finland, pp. 289-299.

"Proceedings of the Offshore Pipeline Technology", OPT European Seminar, 1992.

Krogh, M. A. and Nielson, K. G., "Decade of inspection findings compared with design aspects of two North Sea pipelines", Procs. of The Int. Conf. on Offshore Mechanics and Arctic Engineering, OMAE Vol: 5, 1993, pp. 311-318.

Duncan R. S. et al., "Replacement outfall for Peterhead", J. Inst. Water Environ. Manage., V5 No. 3, June 1991, pp. 265-271.

Karal, K. et al., "Protection, support and installation structure for submarine pipelines", ASME, 1987, pp. 9-16.

Stone, J. H., "The Environmental Impact of a Superport in the Gulf of Mexico", Proc. Nat. Conf. Complete Water Use, 1975, pp. 1217-1229.

"Design Installation and Operation of Pipelines in the Beaufort Sea", Interpipe, 1982, pp. 201-226.

Ekermans, D. and Potgieter D. J., "Design of the pipelines and shore crossings", The Civil Engineer in South Africa, September 1991, pp. 349-356.

Factor, S. and Grove, S. J., "Alaskan Transportation: An Overview of Some Aspects of Transporting Alaskan Crude Oil", Marine Technology, July 1979, pp. 211-224.

Chang, H. H. et al., "Pipeline Shore Approach: Analysis and Design", Presented at the ASCE Fall Convention and Structures Congress, October 25-29, 1982, pp. 853-867.

Hair, J. D., "Pipeline Landfall Construction by Horizontal Drilling", Arctic Offshore Engineering, pp. 792-800.

Schwarz, J. P., "Push-rack method speeds marshland pipelay", Oil & Gas Journal, July 1989, pp. 58-62.

Eddie, A. G. F., "Port Bonython Liquids Terminal, South Australia", First Australasian Port, Harbour & Offshore Engineering Conference, 1986, pp. 44-54.

"The case for the bottom-tow method for laying offshore pipelines", Petroleum Review, November 1976, pp. 593-599.

Martin, D., "Undersea tunnels carry Norwegian 'Pluto' ashore", Tunnels & Tunelling, December 1982, pp. 24-27.

Ivor-Smith, D. J., et al., "Deep soft ground tunnelling under Houston ship channel and Greens Banyon", Proc. Rapid Excavation and Tunnelling Conference, 1989, pp. 229-247.

Ramsey, J. F. et al., "Drilled shore approach sets record on Canadian pipeline", Oil & Gas Journal, October 1991, pp. 72-73.
11

Skonberg, E. R. and Smith, N., "Designing a pipeline installation by horizontal directional drilling", Pipeline & Utilities Construction, June 1993, pp. 14-15.

Black, D. K. et al., "Point Anguello Field trunkline directionally drilled shore crossing", Offshore Technology Conference, 1988, OTC 5742, pp. 509-516.

Peebles, E. E., "Directional drilling to install Arctic marine pipelines", Proceedings of the Conference Arctic '85, Civil Engineering in the Arctic Offshore, pp. 783-791.

Tebboth, L. J., "Dutch field developed with directionally drilled pipe lay", Oil & Gas Journal, April 1987, pp. 59-64.

Fisher, D. L., "Design parameter provide guidance for drilled crossing", Pipe Line Industry, February 1993, pp. 45-49.

Anselmi, A. and Bruschi, R., "Engineering challenges for North Sea pipelines in the nineties", Offshore Northern Seas Conference, 1992.

ACKNOWLEDGEMENTS

The authors sincerely thank Brown & Root management for their support and granting permission to publish this work.

GUIDELINES FOR FREE SPANNING PIPELINES: OUTSTANDING ITEMS AND TECHNOLOGICAL INNOVATIONS

F. TURA
Offshore Division
Snamprogetti S.p.A.
61032 Fano (PS)
Italy

M.B. BRYNDUM
Ports&Offshore Division
Danish Hydraulic Institute
Agern Allé 5
DK 2970 - Hørsholm
Denmark

N. J. R. NIELSEN
Det Norske Veritas
Nyhavn 16
DK1051-København K.
Denmark

ABSTRACT

The aim of the GUDESP project has been to establish internationally recognised Guidelines for the design of free spanning sealines. This was obtained through the development of a unified approach for the determination of the occurrence and the effects of the hydroelastic vibrations and associated load conditions on the structural integrity of suspended spans of offshore pipelines. A key element in this work has been the use of the existing vast experimental data base for the proper scaling of flow induced vibrations. In addition the existing acceptance criteria for free spans have been reviewed by means of a reliability based analysis procedure and partial safety factors are being determined rationalising the design safety levels.

Many technical aspects developed in the GUDESP project are innovative and three main areas are described, namely the structural modelling, the calculation of the response to hydroelastic loads and the calibration of design partial safety factors based on a reliability approach.

INTRODUCTION

The hazard for offshore pipelines may be significant where either irregular seabed or sediment instability and erosions leave suspended pipeline configurations exposed to near-bottom currents. In these cases the free span integrity assessment is of major concern. A number of research and development projects have been performed during the last decade dealing with a number of aspects of pipeline free span assessment. Some examples are the work carried out by J.P. Kenny for the British Department of Energy, the Pipeline Evaluation Manual prepared by Danish Hydraulic Institute, the Submarine

Volume 33: Aspect '94, 59–76.

Vortex Shedding Project and the Methodologies for Adapting Submarine Pipelines to Very Uneven Seabed project undertaken by Snamprogetti.
The design of free spans in submarine pipeline systems incorporates a series of disciplines ranging from structural analysis to environmental engineering, from fatigue analysis to dynamic analysis under extreme and accidental loads. Methods commonly applied when evaluating free spans are briefly summarized.
The starting point for free span assessment is generally the evaluation of the static configuration and the determination of the natural frequencies and the associated modal shapes. Boundary conditions may be more or less realistic particularly in the case of dynamic analysis.
The marine environment is characterised by means of probabilistic criteria based on exceedance curves taken from measured or calculated distributions of currents and waves. The global environmental conditions are then transformed to near pipe conditions, determining the characteristics of the flow and the associated hydrodynamic loads acting on the free span. DNV'81 Rules may be applied although only indications for the hydrodynamic response in steady current are given there. The effects induced by wave action are estimated using different techniques which very much depend on design philosophy and company practice.
The actions of steady and oscillatory flow are generally considered as ideally independent, partly because no indication on how to handle the case exists in the present Rules.
The fatigue life and the resistance to extreme loads are finally performed using the above results.

In practice each company has applied the above procedures and criteria differently dependent on individual Company Standards and project requirements. The complexity of the different procedures ranges from the simple to the very detailed and sophisticated.
The main reasons for the present state of affairs can be summarised:

- the different R&D efforts to address the various problems had a 'fragmented' nature in terms of time, place and goals,
- the physical processes involved in free span dynamics are complex, especially due to the highly non-linear nature of the fluid elastic interaction of the vibrating span and the fluid flow,
- sound and unified rules for scaling experimental data of flow induced vibrations to practical cases covering the full range of environmental conditions are not presently available.

The need for revising, co-ordinating and structuring the assessment of free spans as illustrated by the above was evident when the GUDESP Project was initiated.

THE GUDESP PROJECT

The Development of <u>Gu</u>idelines for the <u>De</u>sign of <u>Sp</u>anning Sealines project was undertaken to produce an internationally recognised Guideline for the evaluation of spans, covering the most practical cases were suspended sections of pipeline are exposed to hydrodynamic loads.
The project started with the collection and the review of the presently applied design

methodologies for free spanning pipelines, among which the candidate methodology was selected.
The major experimental data sources were then identified and the data collected and homogenised in format. Data were furthermore critically re-analyzed and compared in order to establish a consistent and reliable data base applicable for the project.
Some aspects not sufficiently covered by the experimental data were further studied, developing predictive numerical models to be used when drafting the Guideline. A series of numerical models was set-up to check and document the various aspects of the analysis procedure, and a first level approach was drafted allowing the user to make preliminary evaluation.
Finally a software package comprehending the various numerical models was produced, capable of performing advanced free span analysis and the Guideline itself was compiled.
The GUDESP Project has been certified by Det Norske Veritas, who are now involved in the calibration of partial safety coefficients.
Design formats and relevant partial safety factors will be assessed by means of reliability methods; in particular the huge amount of available experimental data will allow to properly consider the probabilistic behaviour of the pipe response to hydroelastic loads in the predictive models.
In a later stage the Guideline will be issued in the form of a DNV document (Certification Note).

GUIDELINE CONTENT

The Guideline has been structured in two parts, the first dealing with the general criteria of analysis and verification of free spans to be applied independently of the degree of sophistication of the models used. The second part gives an example of first level analysis procedure, i.e. the minimum required for the most practical cases not presenting any particular difficulty, of course in line with the concepts expressed in the former.
The principle for the subdivision of the Guideline into two parts is similar to the approach selected in other pipeline rules such as DNV'81 Rules, which contain one part prescribing methods and criteria and another part under the form of Appendices that give methods and calculation procedures in compliance with the general requirements.
The advantage of composing the Guideline of the two parts is that each part deals with a single aspect of the free span evaluation, namely the description of the conceptual basis and the consequent description of calculation procedures.
Part one of the Guideline includes 4 main sections with the following content.

Section 1 describes the main principles, defines the required parameters, units, etc. and gives application areas and exclusions.

Section 2 defines and describes the free span conditions which are required for the evaluations. It is primarily the marine environment and the soil and bathymetry which are of interest. The Guideline includes either specific methods for establishing the required design and evaluation data or acceptable data values are given directly.
This section includes also the principles for free span classification according to which the

span may be classified as isolated or interacting (morphological criterion) and as unevenness or scour induced (temporal criterion).

Section 3 defines the required analysis, and analyses conditions. In principle three different analyses have to be performed:
- static analysis
- dynamic analysis
- fatigue analysis.

These analyses have to be performed for the following free span conditions:
- empty pipe
- waterfilled pipe
- hydrotesting
- operating.

However fatigue calculation should only be performed for free span conditions having durations which may give a notable fatigue damage.
The section also defines the requirements to calculation accuracy and describes acceptable methodologies.
This is described further in the following Chapters of the paper.

Section 4 gives the acceptance criteria to be applied in free span evaluation. The failure types included are:
- yielding of pipe material
- local buckling
- fatigue failure.

The initial considerations regarding failure criteria resulted in adoption of the principles from the DNV 1981 Pipeline Rules.
However safety criteria introduced in other codes of today are generally based on well defined limit states and partial safety factors. It was then decided to apply these principles in GUDESP, but realistic, verified partial safety factors relevant to free spans were not available. It was therefore decided to perform a code calibration using the probabilistic analysis package PROBAN from DnV.
In the following this and a few of the other innovative aspects of the Guideline will be described, namely the structural analysis, the assessment of the hydrodynamic loads and response and the acceptance criteria making use of partial safety factors developed using reliability based methods.

Structural aspects

The modelling of the structural behaviour of spans may in general be performed at different levels of sophistication, depending on aspects considered by the analysis.
The Guideline operates with two basic levels of approach dependent on the particular span conditions. The first level relates to the single span whereas the second level approach is required when free spans are interacting. The morphological criterion of classification thus determines the level of analysis. Figures 1 and 2 are practical means to determine whether spans are isolated or interacting. Considering two spans in sequence

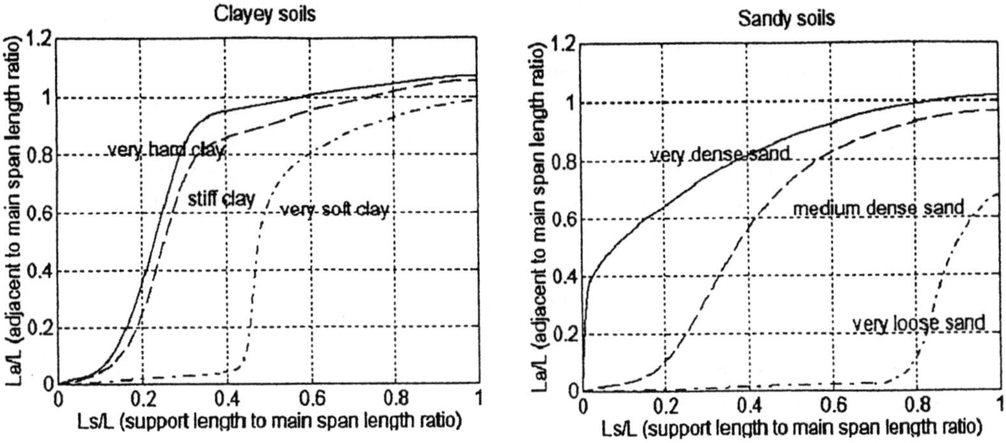

Fig. 1 - Morphological classification of free spans for static calculations

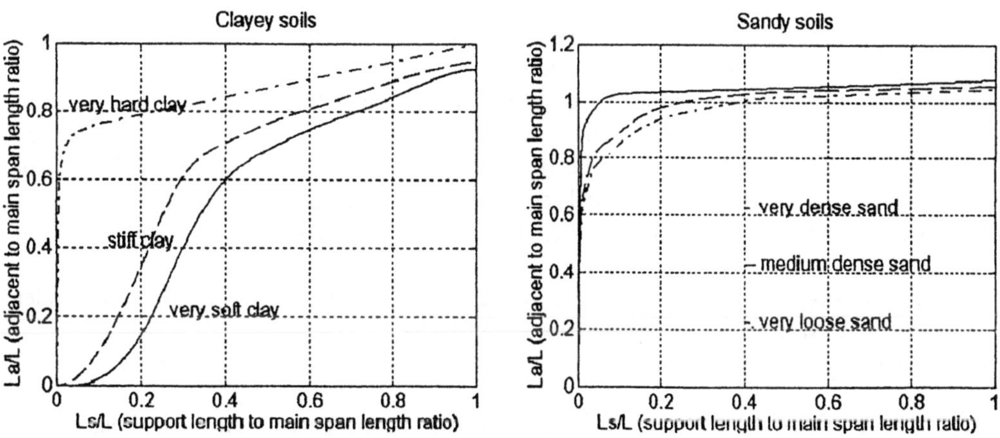

Fig. 2 - Morphological classification of free spans for dynamic calculations

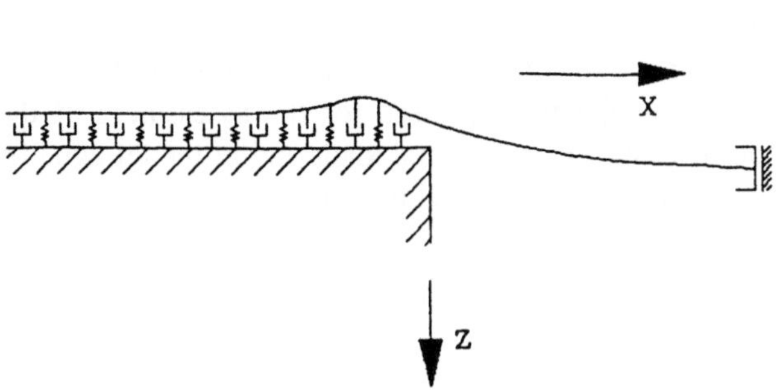

Fig. 3 - Base model for second level procedure

the two ratios adjacent/main span length and support/main span length identify a point in the graphs represented by figures 1 and 2. When such a point falls below the curve corresponding to the encountered soil, the span may be considered as isolated.
When a span is recognised as 'interacting' with other spans, a second level approach for the structural analysis is needed, implementing numerical procedures such as finite element models capable of simulating the soil-pipe interaction including soil elastic-plastic behaviour and the real seabottom profile. Such a procedure has been implemented in the software associated with the Guideline.
When a free span is recognised as isolated, then a first level analysis based on analytical procedures and simplified boundary conditions is sufficient to assess its structural behaviour.
In any case, the analysis of free spans should at minimum include the effects of the following phenomena and conditions:

- soil-pipe static interaction,
- non-linear relation between lateral deflection and axial force due to total or partial axial restraint,
- correct sequence of loading.
  In case of scour induced free span the equilibrium configuration has to be determined in conjunction with the application of all the loads, considering the axial restraint and starting from a rectilinear configuration.
  In case of unevenness induced free spans an intermediate equilibrium configuration has to be determined in conjunction with the empty pipe loads, a constant axial force acting on the pipeline equal to the laying residual axial force and no axial restraint (as-laid configuration). The final equilibrium configuration is to be determined starting from the intermediate one, applying the remaining loads and the axial restraint,
- presence of adjacent spans, where recognised as interacting with the one under consideration (only for the second level approach),
- static configuration when performing dynamic analysis.

An example of first level approach is given in the second part of the Guideline; it was tested against the results given by the second level finite element procedure for a large number of engineering cases and found to give correct results.
The results required from the structural analysis of free span are:

- maximum static stresses,
- dynamic behaviour of the system, characterised by eigenvalues, associated eigenvectors and modal stresses used for the assessment of the maximum dynamic stresses under extreme environmental loads and of the fatigue life,
- maximum dynamic stresses induced by impacting loads, as, for example, those caused by trawl boards.

Second level approach

Acceptable methods will normally have to be based on finite element models and the static equilibrium of a pipe soil system, written for finite elements, is:

$$\left[\mathbf{K}_f + N \cdot \mathbf{K}_g\right]\bar{u} = \bar{F}$$

where :

$\mathbf{K_f}$ = flexural stiffness matrix, resulting from both pipe and soil (where present) contributions,

$\mathbf{K_g}$ = geometric stiffness matrix,

N = T+P, with T = axial force on steel and P = equivalent axial force given by pressure effects,

$\bar{u}$ = equilibrium displacements vector,

$\bar{F}$ = external forces vector, comprising the fictitious effects of pressure and temperature gradient.

The intrinsic non-linearities of the equation above are the soil contribution in $\mathbf{K_f}$ and T, both depending on $\bar{u}$. Moreover, in case significant displacements and rotations, also $\mathbf{K_f}$ and $\mathbf{K_g}$ depend on $\bar{u}$.
A solution procedure of the system of equations above based on simple iterations aiming at updating T, $\bar{u}$ and $\mathbf{K_f}$ and $\mathbf{K_g}$ when necessary is not stable, and divergence of the solutions obtained by iterating is frequently observed, even if relaxation or convergence acceleration methods are applied. The main reason for this is that strong non equilibrated actions may be introduced in the process if the iteration step is too large.
Numerical instabilities may in most cases be avoided by an incremental application of the external loads, that is applying a marching solution instead of a trial and error one.
The reactions induced by the soil to the pipeline can be implemented considering them linearly distributed along the pipe axis, by means of non-linear springs in accordance to the force-displacement law describing the soil elastic-plastic behaviour.
For a given element the constants of the linear distribution of the reactions can easily be determined once the displacements of the nodes limiting the element are known.
Given a displacement v_0, this technique consists in considering a linear spring of intensity $K = P(v_0)/v_0$ where P is the soil bearing capacity. The computational inconvenience appearing when v_0 approaches zero may be solved using limit values for $\dfrac{\partial P}{\partial v_0}$.
Expressions of P are obtained according to consolidated strip foundation theories relevant to cohesive or cohesionless soils, as for example those described in "Rules for the Design Construction and Inspection of Offshore Structures", DNV 1977, Appendix F.
The maximum friction force per unit length of the pipe is simply calculated multiplying the local soil bearing capacity P times the axial friction coefficient f and the local friction modulus is obtained just as the ratio of the local maximum friction force to the axial displacement u of the section. For clays the maximum friction force is upper bounded by the shear strength of the soil.

Pipe-soil interaction in dynamics resembles that of statics, with the only difference that the dynamic reaction of the soil now is considered linear, under the hypothesis of small displacements about the static configuration. Dynamic subgrade moduli are established according to the theories developed for machinery foundations, considering the actual state of local soil deformation when determining soil shear modulus. The axial dynamic reactions are treated as the lateral ones, see, for example, the referenced DNV 1977, Appendix G.

First level approach

In this case a configuration as represented in figure 4 will be sufficiently accurate.
Part two of the Guideline includes a level 1 calculation procedure which gives results in good agreement with the second level procedure for majority of engineering cases.
The soil pipe interaction at span shoulders has been condensed in static and dynamic rotational and axial boundary stiffness, valid for a given set of soils, pipe diameters and span lengths. Such boundary conditions have been calibrated by means of a second level FE procedure and cover a large number of cases. The core of the procedure consists of solving the static case and finding the eigenfrequencies hypothesising the static and modal shapes assumed by the free span. In particular a fictitious length increase factor has been calibrated to be used in the modal analysis, capable of fitting at best the FE results.
The first level procedure is capable to correctly account, in case of isolated free span, for:

- the desired sequence of loads, typical of scouring or unevenness induced free spans,
- the non-linear effect in axial force induced by large lateral displacements,
- the effects of the static deformed shape on the eigenfrequencies.

Hydrodynamic aspects

The hydroelastic vibrations of free spanning pipelines exposed to wave and current action cannot presently be calculated using analytical or numerical methods only. Experimental data, under full or laboratory scale, are essential in the formulation of models predicting the dynamic behaviour of free spans.
The Guideline operates with two different methods or models for predicting hydroelastic vibrations, a response model and a force model. The response model is applied when the vibrations of the free span are dominated by vortex induced resonance phenomena and the force model is used when the free span behaviour is essentially affected by a periodic loading action with no or limited dynamic amplification.
In general it is the flow and the span configuration determining the model to be applied. The review of the experimental data has resulted in the following conclusions regarding the applicability of the models: the response model is applicable for in-line direction in steady current and for cross-flow direction in any flow pattern. The force model is applicable for in-line direction in wave and combined flow.
It has been attempted to develop force models also for the hydroelastic vibration in the cross flow direction. However, due to the complexity of this response particularly in combined flow the resulting force description was at this stage unsuited for direct application in the Guideline.
The assessment of free span fatigue requires in principle information on the magnitude of stress cycles (vibration amplitudes) and the number and distribution of stress cycles. The

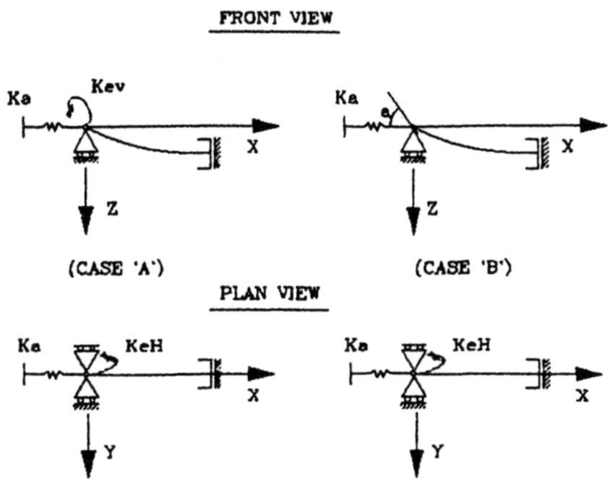

Fig. 4 - Base model for first level procedure

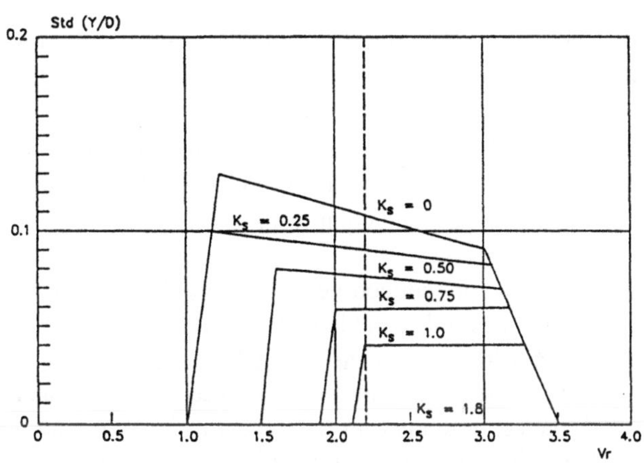

Fig. 5 - In-line response in steady current

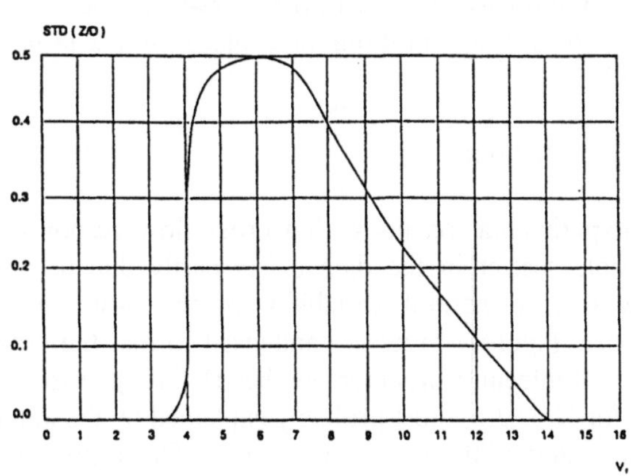

Fig. 6 - Cross-flow response in steady current

ultimate strength of a pipe span is in general associated with the application of an extreme load either in the form of an extreme wave or current induced load.
The hydrodynamic data used in the Guideline are therefore in principle vibration amplitudes and frequencies (response model) and some force coefficients (force model).

Response model - Steady current conditions

Representative steady current test results were available for spring mounted rigid pipe segments, for flexible pipe span models and from full scale measurements. Many parameters have been investigated through the various test programs :

- damping,
- seabed proximity,
- turbulence,
- pipe roughness,
- flow angle of incidence,
- flexibility of pipe span.

The agreement between the different sets of data was not always evident and consequently extensive analyses and evaluations had to be performed to identify the reasons for the discrepancies and for producing idealised response curves (figure 5 and 6). The curves represent a base case characterised by low damping ($K_s \sim 0$), low turbulence ($TI \leq 4\%$), medium pipe roughness (k/D=1/200), a seabottom clearance of e/D≥0.8 and a current direction perpendicular to pipe axis.

In-line vibrations of free pipes for reduced velocities (V_r) between 1.5 and 3.0 caused by vortex shedding have been observed in the laboratory and in full scale.
A number of model tests with pipes near a fixed boundary have displayed a similar behaviour. In other experiments the in-line response has been absent or noticeably reduced with high turbulence, boundary layer effects or high damping.
For Vr > 3.0 the in-line vibrations are associated with the cross-flow vibrations as the main component of the vortex induced forces acts perpendicular to the flow direction. The in-line response is in this case 20-40% of the cross-flow one.
For Vr > 3.0, strong cross-flow vibrations take place with a peak around 5.0 and an evident decay at 11-12.

Response model - Wave conditions

A large number of experimental results giving cross flow response data are available. However a considerable scatter in the data exists partly due to differencies in model characteristics partly due to natural variability of the large number of parameters influencing the response. Figure 7 gives in condensed format the cross flow response in pure wave action. For Keulegan-Carpenter number (KC) equal to 5.0 a particular large response takes place for Vr = 2.5, most probably explained by the initiation of a one sided vortex shedding interacting with the wave motion. The representation of the wave induced cross-flow vibrations together with the ones induced by steady current show a similar behaviour for KC > 10, demonstrating the same basic vortex nature of the

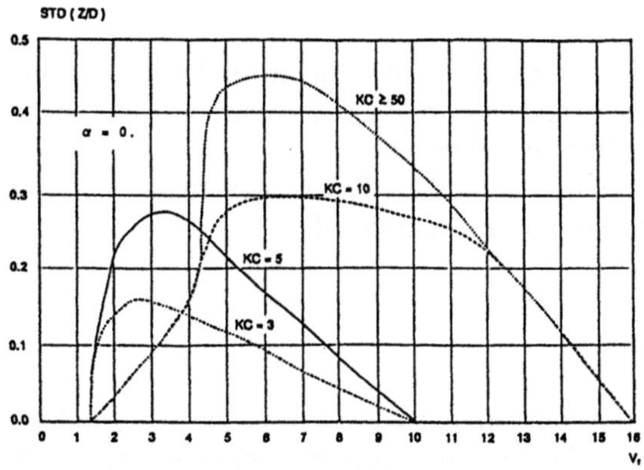

Fig. 7 - Cross-flow response in waves

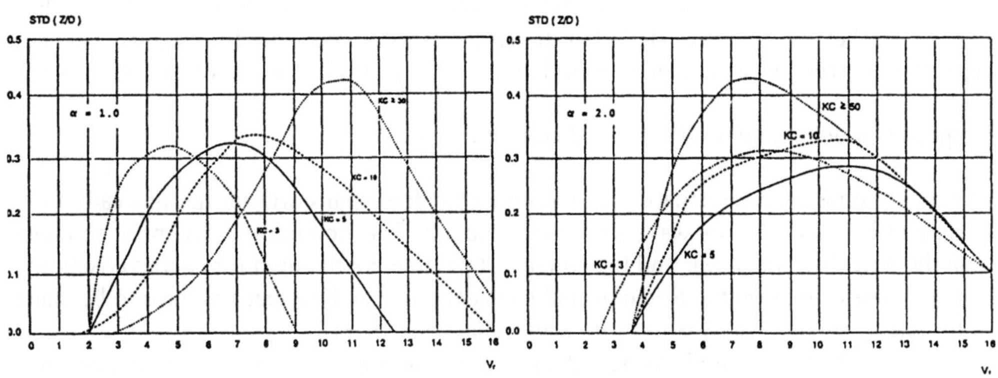

Fig. 8 - Cross-flow response in combined flow

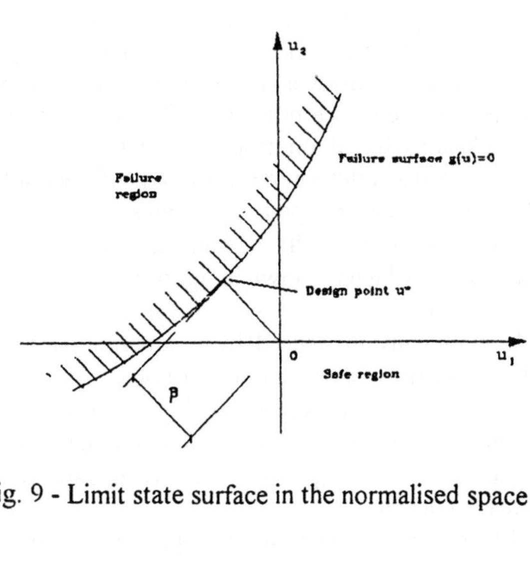

Fig. 9 - Limit state surface in the normalised space

exciting forces.
The implication is that a response model for cross-flow vibrations in waves can be established using the same modalities of the steady current case, including the Keulegan-Carpenter number as additional parameter.
Correction factors for incorporating the effect of governing parameters are given in the Guideline. Following parameters are considered:

- seabed proximity
- pipe roughness
- flow angle of incidence
- damping.

The results obtained from some irregular wave tests showed good agreement with those in regular waves, although in this case it is nearly impossible to separate the influence of the various parameters. The response model established for regular waves will, for the larger KC numbers, result in amplitudes which are larger than those found in irregular waves, primarily because the regular waves give the maximum dynamic amplifications. For the smaller KC numbers ($KC \leq 10$) the large response monitored in regular waves is replicated in irregular waves, demonstrating its validity also in this, more realistic, case.

Response model - Combined wave and current conditions

The most common flow situation of relevance for free spans will be the unsteady flow generated by simultaneous wave and current action. This condition is far more complex than the pure flow situations and the number of parameters required for characterising the flow increases significantly. Some overall considerations and evaluations have to be made to be able to handle this situation with acceptable conciseness.
First, the methodology used for describing the response in combined waves and currents should approach in asymptotic manner the response due to pure flow conditions when either wave or steady current flow becomes dominant.
Further, very little detailed experimental information is available regarding the effect of three-dimensional flow, i.e. the effect of the flow incidence angle and the different directions of waves and current. However, none of the little information available give evidence that applying the independence principle will lead to unconservative results. As a consequence the conclusions based on tests in which waves and currents are co-directional are assumed to be also valid for various oblique flow conditions, provided that only flow components perpendicular to pipe axis are considered.
Also in this case the available test results demonstrate the suitability of a cross-flow response model, an example of which is shown in figure 8.
For reduced velocity below 4 the response is small for small KC numbers and the pure wave response is only marginally influenced by a superimposed current. In general a superimposed current reduces the response in this Vr range.
For KC equal to 5 the effect of superimposing a weak current (α = steady component/orbital component of the flow < 0.5) reduces the response, whereas a stronger current ($\alpha > 2$) increases the response. This may be explained by the approach of the flow situation to steady current.
For KC in the interval from 10 to 25 the observed phenomena are different. At low Vr

values (<4.0) the pure wave response is always larger although only a marginal reduction can be seen when superimposing a steady current. In the intermediate range $4 \leq Vr \leq 8$ the influence of a steady current is negligible. For high reduced velocities ($Vr > 8$) the response in combined flow equals or is higher than that in pure flow.
The calculation of the KC number is based on the oscillatory flow component only, whereas the reduced velocity Vr is calculated using the maximum total flow velocity.

Force model

One of the objectives of this project was to formulate and calibrate a numerical model describing the in-line forces acting on a free spanning pipeline section in pure wave and combined flow. The common industry formulation, the Morison equation, is a priori estimated to be adequate for solving the present problem. However, when resonance occurs due to coincidence of the frequency of the external force with the natural frequency of the excited system, the Morison formulation in its linearised form gives too large response of the free span. In its full formulation the problem of integrating the equations is too time consuming for on-line solution.
A different method has therefore been applied. The basic assumption of this method is that the vibration may be described by the first natural mode satisfying the linear system.
The assumption regarding the shape of the pipe eliminates the need for numerical integration along the span and thereby reduces the computation time significantly.
The coefficients to be applied in the Morison equation are the drag and inertia coefficients. A significant calibration work has been performed in order to determine the most adequate coefficients and their dependence on the main governing non-dimensional parameters.

The resulting recommended force coefficients have the form:

$$C_D = f_{De}(e/D, KC) \cdot f_{Dk}(K/D, KC) \cdot C_D(\alpha, KC) \cdot [1 + S(KC) \cdot STD(Z)/D]$$

$$C_M = f_{mc}(e/D, KC) \cdot C_m(\alpha, KC)$$

where
$f_{De}(e/D, KC)$ and $f_{me}(e/D, KC)$ are the correction factors for seabed proximity effects, $f_{Dk}(k/D, KC)$ is the correction factor for pipe roughness effects,
$S(KC)$ is a corrective function of KC,
STD(Z)/D is the standard deviation of the cross-flow vibration non-dimensionalised with the pipe diameter.
The model was checked against irregular wave test results, verifying that a force formulation gave reasonable and safe results regarding fatigue.

Acceptance criteria and code calibration aspects

The safety philosophy applied in the Guideline is based on partial safety factors including limit states defined by yielding and fatigue of the pipe steel.
In addition a limit state defined by the loss of weight coating integrity due to vibrations induced by vortex shedding is under study.
The calibration of the partial safety coefficients will be carried out utilising the DNV probabilistic program package PROBAN together with the code calibration program package developed within the GUDESP project.
In code calibration the first step is to establish a probabilistic model for determining the reliability of a class of pipeline systems subject to representative load case scenarios. The pipeline systems considered are real cases, representative of the most common conditions met in the offshore industry.
The probabilistic analysis results in a set of reliability measures for the considered design cases allowing to determine the implicit safety levels in the actual design practice. These will be used to assess target safety levels on the basis of which a code calibration procedure is finally applied to determine an optimal set of partial safety coefficients.

The reliability analysis is based upon the formulation of a limit state function, the failure criterion, in terms of load Q and resistance R in the form

$$g(z)=R(z)-Q(z)$$

where $z = (Z_1, Z_2, \dots Z_n)$ denotes the basic variables describing loads, resistance, geometry etc. The limit state function is defined as zero on the limit state surface, as positive for instances in the safe set and as negative for instances in the failure set.
The probability of failure for a given limit state is defined by.

$$P_F = P(g(z) \leq 0) = \int_{g(z)\leq 0} f_z(z)dz$$

where $f_z(z)$ is the probability density function for the basic parameters. Generally the integral above cannot be found analytically and alternative evaluation techniques such as numerical integration or simulation are required.
A generally recognised efficient and sufficiently accurate approximation to P_F is the use of first order reliability method (FORM), see e.g. Madsen, 1986, for a detailed description. The fundamental principle of FORM is to transform the random variables $(Z_1, \dots Z_n)$ into a set of uncorrelated variables $(U_1, \dots U_n)$ which are standardised normally distributed (i.e. zero mean and unit variance). This will map the limit state surface $g(\mathbf{z})=0$ in the **z**-space into a new limit state surface $g(\mathbf{u})=0$ in the **u**-space as illustrated in figure 9. The limit state surface is then approximated by its tangent hyperplane at the design point, $\mathbf{u}^*$, and the minimum distance, β, from origin and to the design point is a measure of the safety margin (first order reliability index). A good approximation to the failure probability can be obtained by

$$P_F \approx \Phi(-\beta) = \Phi\left(-\left|\mathbf{u}^*\right|\right)$$

where $\Phi(\;)$denotes the standard normal distribution function.
While executing the above described procedure additional information is easily obtained regarding the sensitivity measures: importance factors and parametric sensitivity factors. The reliability index vector at the design point can be determined by

$$\mathbf{u}^*=\beta\boldsymbol{\alpha}$$

where $\boldsymbol{\alpha}=(\alpha_1,\alpha_2,...\alpha_n)$ is the unit directional vector. The importance factors are defined by α_i^2 and express the relative importance of the uncertainty of the individual basic variables. The unit directional vector at the design point can be found from

$$\alpha = \frac{\nabla g(\mathbf{u})}{|\nabla g(\mathbf{u})|}$$

in which the gradient vector is defined as

$$\nabla g(\mathbf{u}) = \left(\frac{\partial g}{\partial \mathbf{u}_1}(\mathbf{u}),\ldots,\frac{\partial g}{\partial \mathbf{u}_n}(\mathbf{u})\right).$$

Parametric sensitivity factors are determined by the partial derivative of the reliability index with respect to the basic variables.

A set of reliability indexes (the code realization) is obtained from the probabilistic analysis as a function of the specified code parameters γ_0 (the partial safety coefficients). A measure of closeness between the obtained code realization and the code objective given by a specific target safety level may be expressed by

$$\Delta(\gamma_0) = \int_{-\infty}^{+\infty} f_\beta(\gamma_0)h(\beta)d\beta$$

where $f_\beta(\gamma_0)$ is the frequency density function of β for given code parameters γ_0 and $h(\beta)$ is a weight function. The rationale of the code calibration is to find the partial safety coefficients γ_0 which maximise $\Delta(\gamma_0)$. In case $f_\beta(\gamma_0)$ is a discrete function, a summation replaces the integration above.
A frequent choice of weight function $h(\beta)$ in code calibration is

$$\mathbf{h}(\beta)=-(\beta-\beta_T)^2+\mathbf{p}(\beta)$$

where β_T is the target value of the reliability index and $\mathbf{p}(\beta)$ is a penalty function to ensure conservative results. The use of weight function seeks to minimise the variation of the calculated reliability index for given target safety level.

CONCLUSIONS

Many aspects have been introduced in the Guideline reflecting concepts which are innovative with respect to the current design practice and the actual ruling on the design of free spanning sealines.
Three main topics may be identified:

- structural modelling,
- hydroelastic loads/response,
- rational design formats.

As regards structural modelling, emphasis has been put in relating the kind of design approach to the kind of free span. To this purpose two classification criteria have been introduced, namely the morphological and the temporal criteria.
The first one aims at defining the free spans as isolated or interacting, depending on the degree its static and/or dynamic behaviour is affected by the presence of other adjacent spans. The studies performed allowed to produce a first approach method to determine whether a span is isolated, to be used for static and dynamic interaction assessment, where sophisticated numerical models are usually required to check interacting spans.
The temporal criterion aims at determining the time of occurrence of the span. Typically an uneven seabed gives rise to spans just after laying operations, while an erodible seabed may cause free spans to occur at any time during the pipeline operating life.
The time of occurrence of a free span determines the sequence of loads to consider in the static structural analysis.
The seabed modelling has been treated at two levels, the first one numerically, describing the pipe-soil sectional behaviour, the second one giving elastic properties condensed at span ends, to be used when applying simplified procedures.
As regards hydroelastic loads and responses the Guideline reflects the most recent findings of many research projects, condensing the evidences coming from a huge amount of experimental tests.
All kinds of flow are treated, ranging from purely steady to purely wavy conditions; the most suitable methods to assess the free span responses are indicated, whether load or response based.
In particular the existence of coupling between vertical and horizontal oscillations is stated in the typical cross-flow synchronisation region, taking into account the 3-D structural behaviour of real spans. Moreover, the event of cross-flow vibrations is presented for any combination of steady current and wave.
As regards rational design formats these are actually under development. They will be based on partial safety factors to be obtained by means of a reliability analysis performed calibrating the code package developed within the project.
The limit states which will be included are defined by yielding and fatigue of the pipe steel. The possibility of considering a limit state defined by the loss of weight coating integrity due to vibrations induced by vortex shedding is under study.

ACKNOWLEDGEMENTS

The authors wish to acknowledge the support to the project given by the EC Authorities, Exxon Production Research, Statoil and SNAM S.p.A.. Please note that the text reflects the authors' personal opinions, which not necessarily correspond to those of the cited Companies.

REFERENCES

Bernetti, R. et alii (1990), "Theoretical and Experimental Analysis of Soil-Pipe Interaction at Free-Span Shoulders for Oscillating Pipelines", European Offshore Mechanics Symposium, Trondheim.
Blevins, R. D. (1990), Flow-Induced Vibrations, Van Nostrand-Reinhold.
Bowles, J. E. (1988), Foundations Analysis and Design, McGraw-Hill.
Bryndum, M.B. et alii (1989), "Long Free Spans Exposed to Wave Flow and Steady Current", Offshore Technology Conference, Houston.
Bruschi, R. et alii (1986), "Scour Induced Free Spans", OMAE Conference, Tokio.
Bruschi, R. et alii (1988), "Submarine Pipeline Design Against Hydroelastic Oscillations: the SVS Project"; OMAE Conference, Houston.
Bruschi, R. et alii (1989), "Field Tests with Pipeline Free Spans Exposed to Wave Flow and Steady Current", Offshore Technology Conference, Houston.
Bruschi, R. et alii (1991), "Free Spanning Pipelines: A Review", ISOPE Conference, Edinburgh.
Bruschi, R. and Vitali, L. (1991), "Large-Amplitude Oscillations of Geometrically Nonlinear Elastic Beam Subjected to Hydrodynamic Excitation", Journal of Offshore Mechanics and Arctic Engineering, Volume 113.
Bruschi, R. et alii (1993), "The MASPUS Project, Methodologies for Adapting Submarine Pipelines to very Uneven Seabed", OMAE Conference, Glasgow.
Danish Hydraulic Institute (DHI) and Ramboll and Hannemann (1985), Danish Submarine Guidelines, Danish Energy Agency.
Dean, R.G. and Dalrymple R.A. (1983), Water Wave Mechanics, Prentice Hall.
U.K. Department of Energy (1989), Submarine Pipelines Guidance Note.
Det Norke Veritas (1977), "Rules for the Design Construction and Inspection of Offshore Structures".
Det Norke Veritas (1981), Rules for SubmarinePipeline Systems.
Det Norske Veritas Sesam A/S, General Purpose PROBabilistic ANalysis program (PROBAN)
Griffin, O.M. et alii (1973), "Measurements of the Response of Bluff Cylinders to Flow-Induced Vortex Shedding", Offshore Technology Conference, Houston.
Griffin, O.M. et alii (1975), "The Resonant, Vortex-Excited Vibrations of Structures and cable System", Offshore Technology Conference, Houston.
Huang, Y.M. et alii (1987), "Nonlinear Response of a Dynamic System Due to Oscillatory Flow", *Journal of Offshore Mechanics and Arctic Engineering, Vol. 109.*
Jacobsen, V. et al. (1984), "Vibrations of Offshore Pipelines Exposed to Current and Wawe Actions", OMAE Conference, New Orleans.

King, R. (1977), "A Review of Vortex Shedding Research and its Applications", *Ocean Engineering, Vol. 4, pp. 141-172.*
Madsen, H.O. et al. (1986), Methods of Structural Safety, Prentice Hall Inc., Englewood Cliffs.
Morison, J.R. (1950), "The Force Exerted by Surface Waves on Piles", AIME Petroleum Transactions, Vol. 189.
Palmer, A. (1990), "Pipeline Spans", Course on Offshore Pipeline Engineering, London.
Palmer, A.C. and Kaye, D. (1991), "Rational Assessment Criteria for Pipeline Spans", Offshore Pipeline Technology, Copenhagen.
Orgill, G. et alii (1990), "Effects of Structural Boundary Conditions on Fatigue Life of Pipeline Spans Undergoing Vortex-Induced Vibrations", OMAE Conference, Houston.
Raven, P.W.G. (1986), The Development of Guidelines for the Assessment of Submarine Pipeline Spans, UK Dept. of Energy, OTH 86 231.
Sarpkaya, T. (1976), "Forces on a Cylinder Near a Plane Boundary in Sinuscidally Oscillating Flow", *Journal of Fluids Engineering, Vol. 98.*
Sarpkaya, T. (1979), "Vortex Induced Oscillations", *Journal of Applied Mechanics, Vol. 46.*
Sarpkaya, T. and Rajabi, F. (1979), "Hydrodynamic Drag on Bottom Mounted Smooth and Rough Cylinders in Periodic Flow", Offshore Technology Conference, Houston.
Sarpkaya, T. and Storm, M. (1985), "In-line Force on a Cylinder Translating in Oscillatory Flow", *Applied Ocean Research, Vol. 7, n. 4.*
Tsahalis, D.T. (1985), "Vortex-Induced Vibrations Due to Steady and Wave-Induced Currents of a Flexible Cylinder Near a Plane Boundary", OMAE Conference.
Tura, F. et alii (1993), "A Simplified Method for Assessing the 3-D Structural Behaviour of Free Spanning Sealines Exposed to Steady Currents", EURODYN'93, Trondheim.
Tura, F. et alii (1990), "Non Linear Behaviour of Free Spanning Pipelines Exposed to Steady Currents: Model Tests and Numerical Simulations", European Offshore Mechanics Symposium, 1990.
Vitali, L. et alii (1993), "Dynamic Excitation of Offshore Pipelines Resting on Very Uneven Seabeds", Eurodyn, Trondheim.
Williamson, C.H.K. (1985), "In-line Response of a Cylinder in Oscillating Flow", Applied Ocean Research, Vol. 7, n. 2.
Wooton, R. et alii (1978), "Oscillation of Piles in Marine Structures. A descritpion of the Full-Scale Experiments at Immingham", CIRIA Tech. Note 40.

UPHEAVAL BUCKLING OF FLEXIBLE PIPES METHOD SELECTED AT THE TROLL OLJE FIELD

S.A. LØTVEIT
SeaFlex a.s
P.O.Box 451, 1371 Asker
Norway

P. BRYN and B.K. HJERMANN
Norsk Hydro a.s
P.O.Box 200, 1321 Stabekk
Norway

1. ABSTRACT

Protection of flexible pipes by rock dumping or trenching has been performed on several North Sea projects. Numerous projects have experienced that flexible pipes are difficult to stabilise. Flexible pipes have been reported to break through the protection cover in both upheaval and lateral buckles. Such problems can be related to the pipe design, vertical imperfections, installation method, protection method, protection design or an unfavourable combinations of such parameters. Protection of flexible pipes is required in several applications both as thermal insulation and as impact protection.

The behaviour of protected flexible pipes differs from conventional steel pipes. The main differences are:

- The bending stiffness is low and non-linear.
- The axial stiffness is non-linear.
- The pressure and temperature induced elongation are design dependent and may be significantly larger than for a steel pipe.
- The pipes are normally installed with low back tension and are normally not straight after laying.
- Rock-dumping or trenching while the flexible pipe are pressurised (pre-stretched) has been performed. However, this method has not solved the break through problem.

Volume 33: Aspect '94, 77–110.

The Norsk Hydro Troll Olje field will be a major user of protected flexible pipes. The selected protection for Troll Olje is tailor made for the actual flexible pipes and allows controlled lateral deflection. The selected solution includes the following protection methods:

- Trenching with natural backfill
- Trenching and rock-dumping
- Rock-dumping only

This paper is organised as follows:

- Section 2 presents the 70 km of flexible flowlines which will be installed at the Norsk Hydro Troll Olje field.
- Section 3 presents the flowline protection requirements related to impact, fishing, anchor chain crossing and thermal insulation.
- Section 4 presents the theoretical basis for flexible flowline design to avoid upheaval buckling.
- Section 5 covers the calculated behaviour of a flexible flowline, trenched and covered with natural backfill.
- Section 6 covers the calculated behaviour of a trenched pipe with rock dump cover.
- Section 7 summarises the protection methods selected for the Troll Olje field and presents why these methods are considered the optimum for this field development.

After the first installation campaign is finished in 1994 a detail evaluation of the selected protection method will be performed to confirm that the selected method is adequate. Alternatively, make corrections as required.

2. TROLL FIELD DESCRIPTION

2.1 Field Description

The Troll Field is located approximately 70 km west of Bergen with water depth ranging from 300 - 340 m. It consists of the Troll East and the Troll West sections and is contained within blocks 31/2, 31/3, 31/5 and 31/6. The Troll West sections are further divided into the oil province (TWOP) and the gas province (TWGP).

Norsk Hydro is the operator for the development of the oil province and is further planning a stage by stage development of the oil reserves in the gas province.

The oil province will be developed by horizontal subsea wells drilled from 18 subsea satellite wells. The subsea wells are grouped in four clusters. Each subsea well is tied back to one of the four manifold centres through 6" and 8" Wellstream flexible flowlines and integrated service umbilicals (ISU). The flexible flowlines and ISUs are typically 160 to 360 m long. In addition to the short flowlines and ISUs, two remote satellite wells are tied back through approximately 1500 m long flowlines and ISUs.

From each manifold centre two 10" (Coflexip) or two 9" (Wellstream) flexible production lines, denoted gathering lines and one ISU will be tied back to a floating production platform via a riser base and flexible risers. The length of the gathering lines ranges from 3.5 km to 9.5 km. In addition, a 4 km long 10" (Coflexip) gas injection line is routed directly from the riser base to a gas injection well.

The oil will be exported to shore through a pipeline. The associated gas will be exported to the Norske Shell Troll phase 1 GBS platform.

The following pressures and temperatures are applicable for the f lexible lines:

Condition	Temperature	Pressure
Normal Production	58 - 63°C	45 - 90 bar
Design	68°C	178 bar
Hydraulic Test	-	223 bar
Trenching	-	245 bar

TABLE 2.1

The trenching pressure is higher than the design pressure to account for the thermal expansion.

Due to installation sequence and positioning of subsea valves the gathering lines will be pressurised and depressurised several times after the pipes have been installed and protected.

The general layout of the Troll Olje development is illustrated in figure 2.1.

2.2 Installation

The installation of the flexible lines for the Troll Olje project will start in summer 1994 and continue through the 1995, 1996 and 1997 season. All gathering lines, the gas injection line and 9 of the flowlines and the associated ISUs are planned to be installed and protected before production start in January 1996.

The remaining flowlines and ISUs will be installed during 1996 and 1997 following completion of the corresponding production wells.

Year of Installation	Total Gathering Line Length (no. of lines)	Total Length, Flowline/Gas Injection Line (no. of lines)	Total ISU Length (no. of lines)
1994	14.8 km (4)	-	7.4 km (2)
1995	28.8 km (4)	6.5 km (9)	17.6 km (13)
1996	-	2.5 km (6)	2.5 km (6)
1997	-	1.0 km (4)	1.0 km (4)

TABLE 2.2

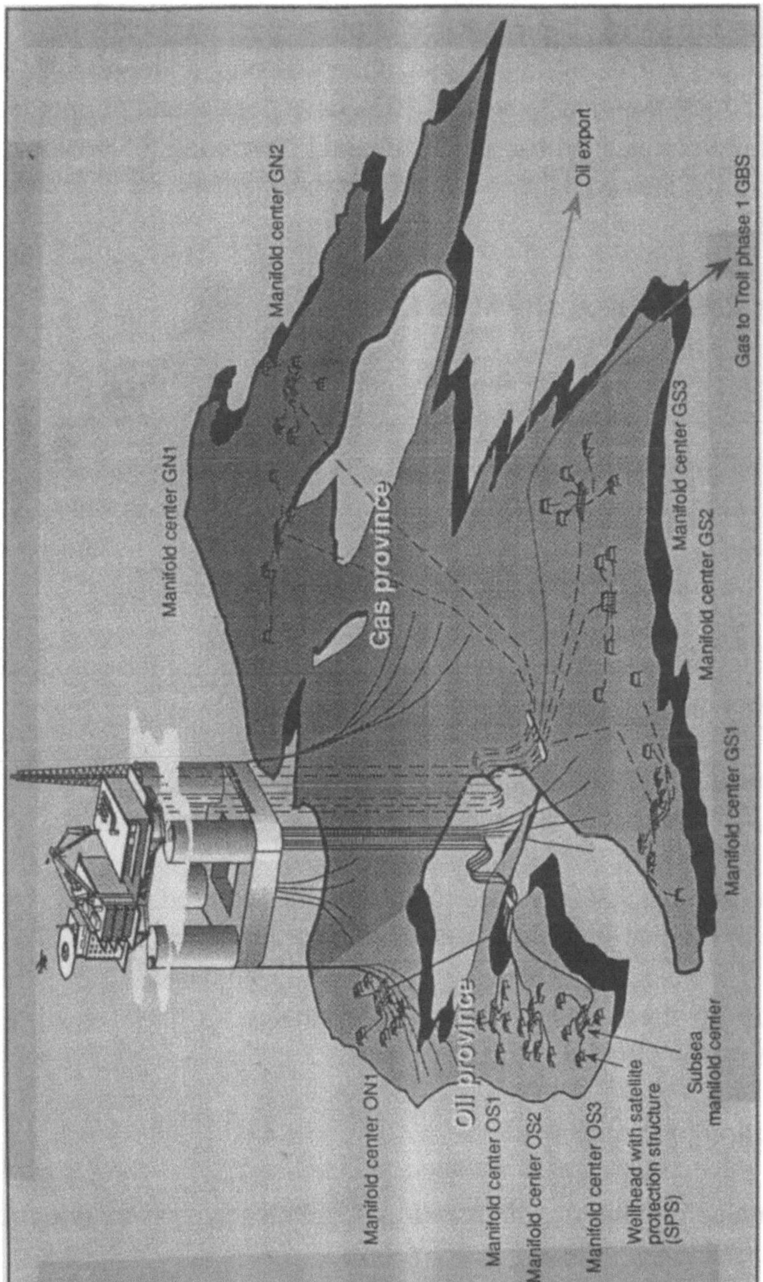

FIGURE 2.1

TROLL FIELD LAYOUT

2.3 Seabed and Soil Conditions

The seabed soil in the Troll area consists of soft clay with a typical undrained shear strength of 4 kPa. A high density of seafloor depressions or pockmarks is mapped in the Troll area. Typical diameter of the pockmarks are 50 - 100 m and typical depth is 5 - 8 m. In addition numerous smaller depressions with typical depth of 1 m and typical width of 10 m have been found during detail mapping. By carefully planned subsea layout and line routing the pockmarks are avoided and the lines are laid on a relatively flat seabed.

3. PROTECTION REQUIREMENTS

The following protection requirements are applicable for the Troll Olje subsea lines:

- The lines with protection shall be over trawlable. This means that no hooking is accepted and the lines when protected shall withstand 10 kJ impact or be protected such that impact will be avoided.

- In the cluster areas the subsea lines shall be protected against dropped objects from drilling rigs and installation activity (50 kJ).

- In the route approach to the cluster areas the subsea lines shall be protected from anchor chain sweep and anchor chain touch-down.

- Insulation from the soil is needed to achieve the required thermal insulation of the two long flowlines and the gathering lines.

The following protection methods have been evaluated for the Troll Olje gathering lines and the long flowlines:

- No protection of the lines.

- Rock dumping of gathering lines and ISU laid closely spaced, one rock-berm required.

- Rock dumping of individual lines laid 20 m apart.
- Trenching of gathering lines and ISU, laid 20 m apart. Natural soil backfilling during trenching. Rock dumping of trenched lines in the areas where dropped object and anchor chain protection is required.
- Local use of mattresses and protection covers close to structures.

For the short lines the following methods have been evaluated:

- Rock dumping of short flowlines and ISUs.
- Combined trenching and rock dumping of the short flowlines and ISUs.

4. FLEXIBLE PIPE PROPERTIES

4.1 General

This section presents the properties of flexible pipes and discusses why flexible pipes have larger buckling potential than steel pipes.

Flexible pipes are composed of a polymeric layer which acts as a pressure barrier and several reinforcement layers. Bending flexibility is achieved due to slip between the individual wires in the reinforcement layers. Figure 4.1. shows a typical pipe design.

The flexible pipe manufacturer has several options to tailor make a design for a specific application. With respect to upheaval buckling the two most important parameters are:

- Pressure and temperature induced elongation.
- Bending stiffness.

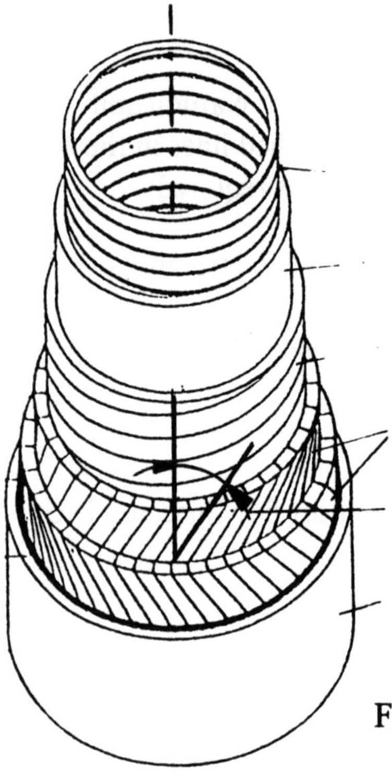

FIGURE 4.1

4.2 Pressure and Temperature Induced Elongation

The driving force for an upheaval buckle is the pipe elongation. Flexible pipes with no pressure spiral may be tuned to have negative pressure induced elongation, zero axial elongation or positive axial elongation. This is achieved by change of lay angle. Zero pressure induced elongation is achieved with a lay angle of 55 deg, (neutral angle). With armouring at this angle the radial expansion balances the axial elongation.

In a high pressure application additional hoop reinforcement is normally used to reduce armour stress and to reduce radial deformation. In a non-bonded pipe an interlocked pressure spiral is used. By inclusion of this spiral the radial expansion is reduced and pressure induced axial elongation can not be prevented.

In a design with pressure spiral and the axial stiffness of the pipe increases with small lay angle of the armouring. With an armouring lay angle of zero degrees, the pressure induced axial elongation of a straight pipe is at a minimum. To ensure controlled bending behaviour the lay angle should not be less than approximately 25 deg.

For a steel pipe the hoop strain will generate a negative axial strain component due to the Poisson effect. This effect reduces the pressure induced axial strain with 60%. In a high pressure flexible pipe this effect is not present. In addition, flexible pipes use more high strength steel qualities resulting in larger permissible strains, typically 50% larger than for steel pipes.

The temperature induced axial elongation is not influenced by material strength and the Poisson effect. The temperature induced axial elongation of steel pipes and flexible pipes will hence be comparable if the wall temperature is the same and there are no gaps between the pressure spiral and the tension armours.

To conclude, high pressure flexible pipes have significantly larger pressure induced axial expansion than steel pipes. The magnitude of the temperature induced axial elongation is approximate the same for flexible pipes and for steel pipes.

4.3 Bending Stiffness

Flexible pipes are designed with as large bending flexibility as possible. This flexibility is obtained by use of flexible polymeric tube layers and thin reinforcement wires.

The wires slides relative to each other when the pipe is bent. For a flexible pipe with internal friction the initial bending stiffness will be influenced by friction. When the static friction is exceeded the slope of the bending moment curve is close to the bending stiffness for the polymeric tubes only, see figure 4.2.

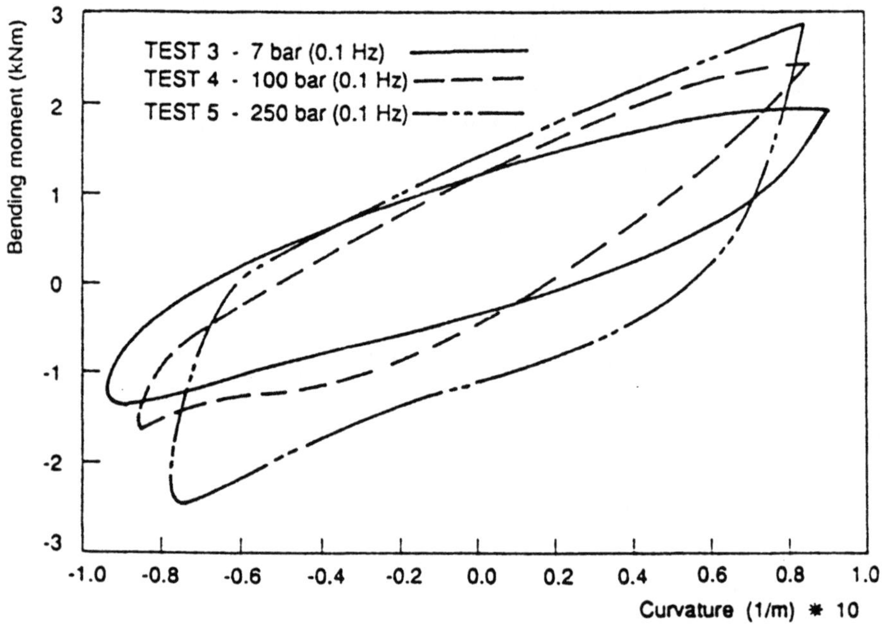

BENDING MOMENT CURVE FOR FLEXIBLE PIPES /3

FIGURE 4.2

For an axially compressed pipe laid on the seabed the following behaviour is expected:

- When the axial compressive load is applied the pipe deflects laterally to its theoretical shape which is governed by lateral and axial friction, pipe length, axial stiffness and bending stiffness.

- After the pipe has remained in this position for some time, the bending stresses in the pipe wall decays due to shake down of friction effects and creep of the polymer.

- In subsequent reduction and re-application of the axial compression (shut-down followed by re-pressurisation) the pipe will behave as if the pipe has a permanent bend.

4.4 Buckling of Flexible Pipes

As discussed in section 4.2 the axial elongation of high pressure flexible pipes will be greater than for steel pipes. The bending stiffness is low and the bending moment resulting from curving the flexible pipe will decay with time.

Pressurisation of unprotected flexible pipes will result in lateral deflection. The deflection "wave-length" will be short compared to steel pipes. Short "wave-lengths" "absorb" more axial elongation than long wave lengths. As a result the lateral deflection amplitude will be moderate.

Figure 4.3 shows a calculated shape of a laterally deflected flexible pipe, laid on the seabed and pressurised. The deflection amplitude and length depend on pipe stiffness, elongation, friction and "as laid" route. Figure 4.3 is a large amplitude deflection for the gathering lines. The pipe will be trenched and rock dumped in this pressurised mode.

The calculated force required to initiate the deflection was 65 kN. This force depend on as laid straightness, 100 kN is considered a maximum value. The axial compression when the deflection is fully developed is approximately 8 kN. The feed-in length will be up to 1000 m on each side of the buckle.

The required lateral resistance to avoid that the pipe straighten out when the pressure is bled off may be estimated as follows:

$$R_{Lat} = T/R_c \qquad (4 - 1)$$

where

R_{Lat} = Lateral resistance

T = $EA \cdot \varepsilon_{ax}$

EA = Pipe axial stiffness

ε_{ax} = Pipe axial elongation due to internal pressure

R_c = Radius of curvature in buckle

Based on equation 4 - 1 it can be seen that the required lateral resistance to hold a flexible pipe in position will be larger than for steel pipes as R_c will be smaller and T will be approximately the same.

Norsk Hydro has specified that the flexible flowlines and gathering lines should be designed to minimise the buckling potential. Due to the service requirements the manufacturers selected pipe designs with pressure spiral.

As discussed above a flexible pipe with pressure spiral will always get a significant pressure induced axial elongation. The behaviour of the flexible pipes in a trench with natural backfill and in a trench with rock cover is evaluated in detail to optimise the protection method.

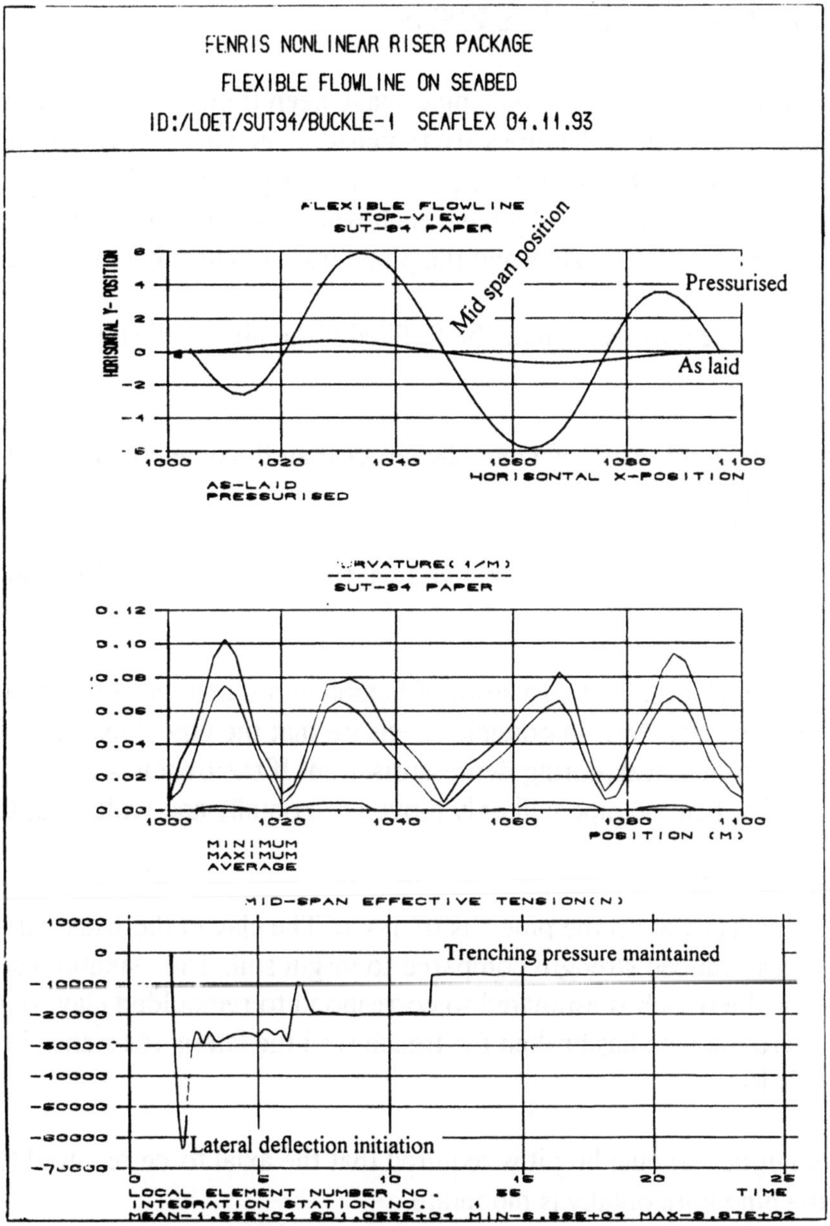

NON-LINEAR FENRIS ANALYSIS /2/ OF GATHERING LINE PRESSURISED BEFORE TRENCHING

FIGURE 4.3

5. TRENCHED FLEXIBLE PIPES

In recent offshore projects flexible pipes have been trenched in pressurised mode. With this method the pipes are trenched in an axial elongated state. When the pressure is bled-off the pipe tries to straighten out. This straightening is counteracted by the trench wall. A balance is obtained between trench wall resistance and the pipe tension induced contact load.

The following requirements must be fulfilled if the pipe shall stay in the trench:

- The soil lateral resistance must be larger than the pipe contact load, see equation (4 - 1).
- The trench wall must be nearly vertical to prevent the pipe from sliding out.

Due to the small radius of curvature of flexible pipes and the soft clay at the Troll field it is, however, impossible to ensure that the pipe does not break through the trench wall during depressurisation. If the trench is deep with vertical walls then the pipe will only move horizontally and widen out the trench.

During repressurisation the pipe has to move. The clay in the widened out trench has a reduced strength compared to virgin soil. The maximum soil strength in the trench is estimated to correspond to remoulded clay. In the areas where the pipe has broken the trench wall the lateral resistance behind the pipe is low.

To avoid upheaval buckling it is required that the axial force required to move the pipe horizontally is moderate.

Analyses have been performed with the non-linear program FENRIS /2/. The following effects have been considered:

- Bending stress relaxation with time. In the calculation (figure 5.1) the pipe is stress free in the curved as-trenched configuration.

- Pressure and temperature induced elongation.

- Lateral soil strength modelled with virgin properties during initial penetration and remoulded strength when moving in the widened trench. Figure 5.2 shows the soil model.

- Vertical uplift resistance considering pipe weight and overburden due to natural backfill. Figure 5.3 shows the vertical uplift resistance.

The scenario considered is 2 km of 10" flexible pipe which is 0.2% elongated. The pipe has been given a theoretical lateral deflection, cosine shape with 50 m length and 10 m double amplitude. This deflection length is relatively long for flexible pipes.

As shown in table 5.1 and figure 5.1 the lateral soil resistance is not sufficient to avoid that the pipe break the trench wall during depressurisation. The trench is widened out. The tension in the pipe is 120 kN when the trench sidewall breaks. The tension is increasing steadily as the pipe is moving into virgin clay. To clearly demonstrate the response in a trench with natural backfill the analyses shown in figure 5.1 are based on long feed-in length and little axial friction. The lateral motion during depressurisation and in particular during repressurisation are hence amplified. The general behaviour is, however, properly represented.

Figure 5.3 shows the results of a traditional upheaval buckling calculation The sensitivity to uplift resistance (w), axial friction (my) and pipe bending stiffness (EI) is shown in the figure. The minimum calculated axial force to generate an upheaval buckle is approximately 200 kN. With this axial compression an upheaval buckle will take place if there is no natural backfill.

The compression force initiating the lateral deflections was estimated to 60 kN. The lateral deflections assumed during trenching are relatively long, longer than calculated for the as laid line, see figure 4.1. Shorter lateral deflections during trenching will reduce the required force to initiate lateral deflections. The initiation force for lateral deflections is hence smaller than the force required to generate upheaval buckles.

Note that natural backfill is in this case not required to prevent upheaval buckles.

The following conclusions apply for the 10" gathering lines trenched with natural backfill.

- The pipe should be trenched pressurised with lateral deflections.

- During depressurisation it is expected that the pipe will break the trench sidewall. The pipe tension when the sidewall breaks will typically be 50 - 200 kN.

- Repressurisation will result in new lateral deflections in the area where the pipe moved during depressurisation. The length and amplitude of these buckles will be significantly less than the deflection "wave-length" of the pipe before trenching.

- Upheaval buckling will not take place. Even with no natural backfill the axial force required to generate an upheaval buckle is 3 times larger than the force required to generate lateral deflections.

Input	Pipe outside diameter	(m)	0.345
	Pipe bending stiffness	(kNm^2)	100
	Pipe axial stiffness	(kN)	600,000
	Pressure induced axial force for a pipe restrained from any displacement	(kN)	600
	Pipe elongation during trenching	%	0.2*
	Pipe minimum radius of curvature during trenching	(m)	12.7
Depres-surisation	Pipe tension when trench side starts to break	(kN)	120
	Pipe maximum lateral displacement during depressurisation	(m)	3.5
	Pipe tension after depressurisation	(kN)	340
	Pipe minimum radius of curvature after depressurisation	(m)	~ 20
Repres-surisation	Pipe lateral displacement after repressurisation	(m)	1.4**
	Pipe vertical motion during repressurisation	(m)	0
	Pipe axial compression during initiation of lateral deflection	(kN)	60
	Pipe axial compression in fully developed lateral deflection	(kN)	40
	Pipe minimum radius of curvature after repressurisation	(m)	5.9

* Model includes 1000 m straight pipe with no axial friction on each end of buckle.

** Not in same position as displacement during depressurisation, see fig 5.1

TRENCHING WITH NATURAL BACKFILL, ANALYSIS KEY RESULTS

TABLE 5.1

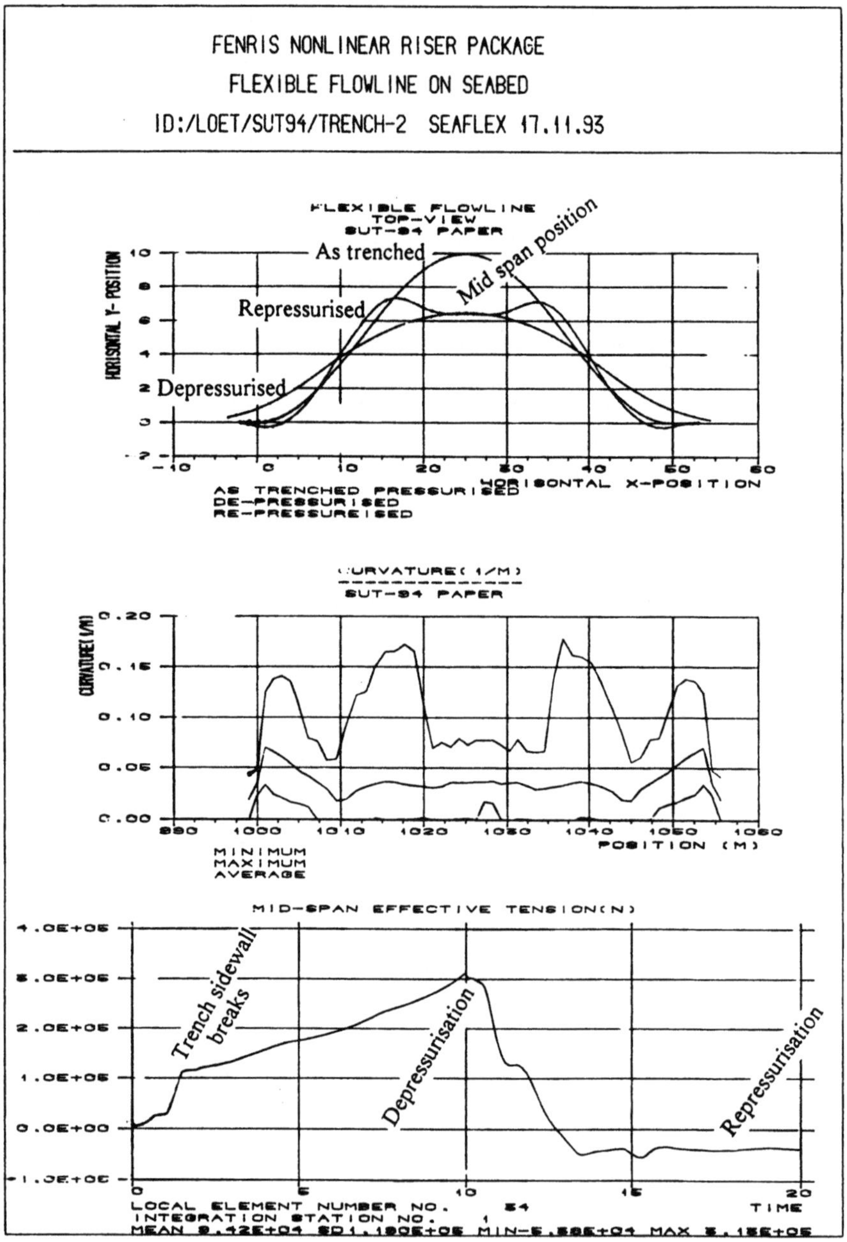

NON-LINEAR FENRIS ANALYSIS OF 10" PIPE FLEXIBLE FLOWLINE TRENCH WITH NATURAL BACKFILL PRESSURISED BEFORE TRENCHING

FIGURE 5.1

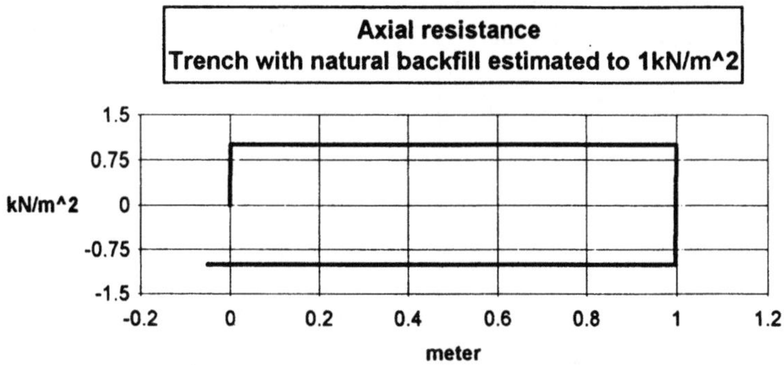

a) AXIAL SOIL RESISTANCE

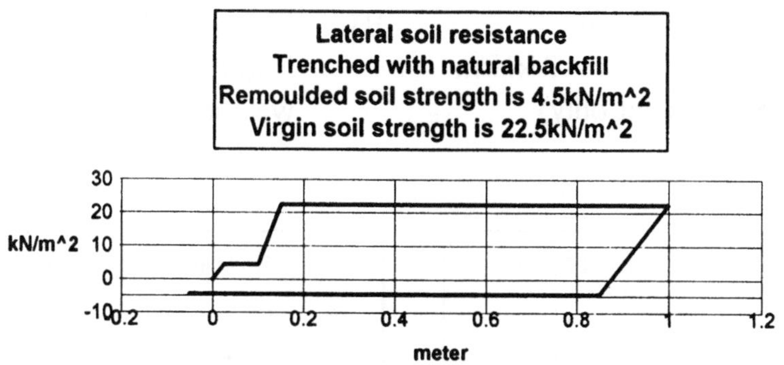

b) LATERAL SOIL RESISTANCE

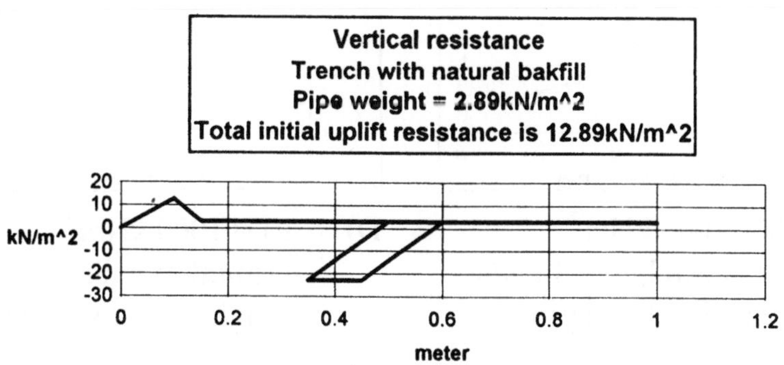

c) VERTICAL SOIL RESISTANCE

TRENCH WITH NATURAL BACKFILL
ELASTOPLASTIC MATERIAL USED TO MODEL SOIL PROPERTIES

FIGURE 5.2

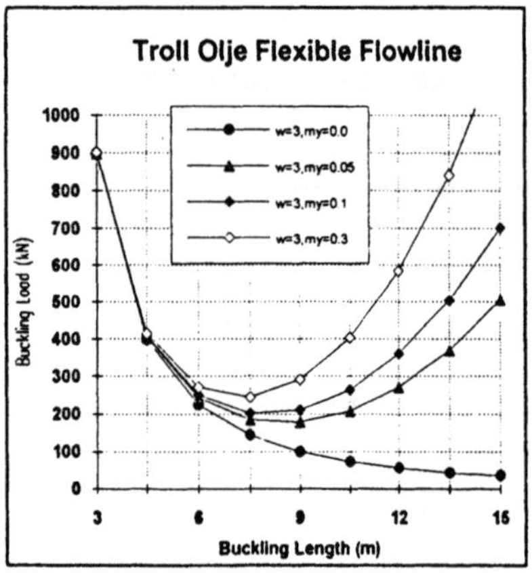

A) EI = 100 kNm^2
No backfill, w = 3 kN/m
(pipe weight only)

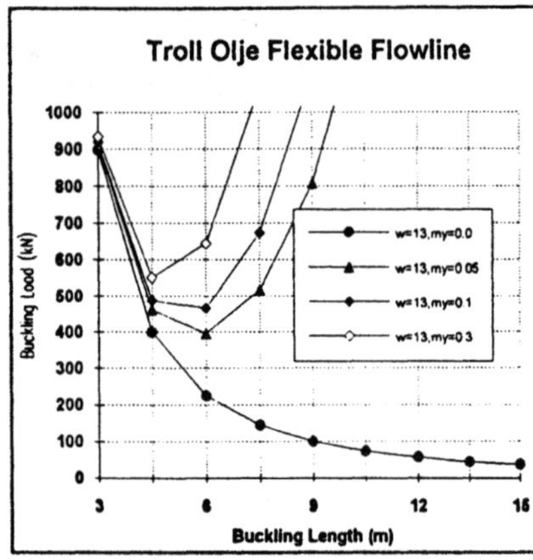

B) EI = 100 kNm^2
Natural backfill and pipe weight, w = 13 kN/m^2

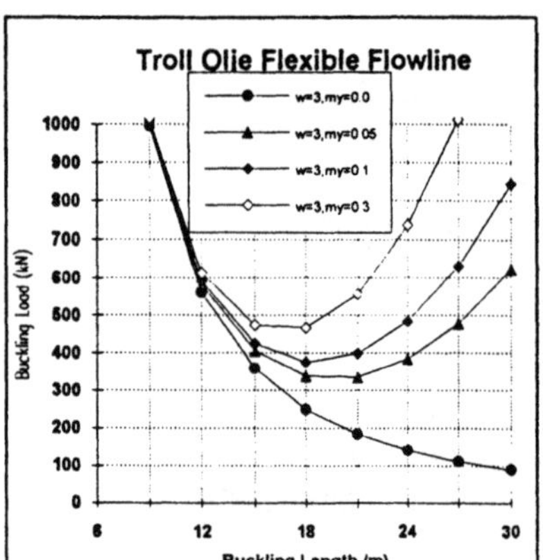

C) EI = 1000 kNm^2
No backfill, w = 3 kN/m
(pipe weight only)

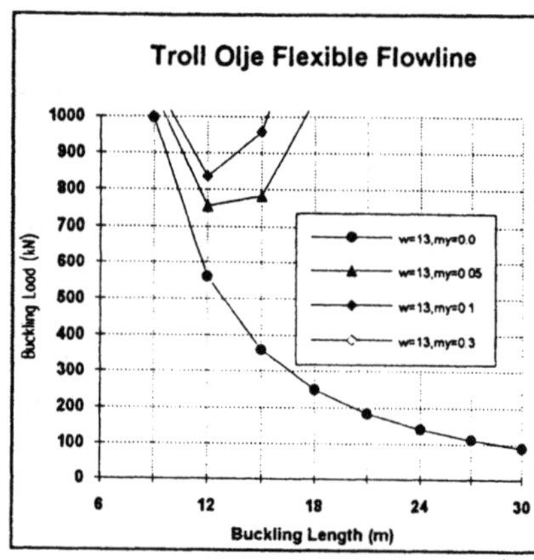

D) EI = 1000 kNm^2
Natural backfill and pipe weight, w = 13 kN/m^2

UPHEAVAL BUCKLING CALCULATIONS BASED ON TRADITIONAL CALCULATION METHOD /1/

FIGURE 5.3

6. ROCK DUMPED FLEXIBLE PIPES

As for the trenched flexible pipe a stable configuration may be achieved if the pipes are rock dumped at maximum elongation and the lateral strength is sufficient to avoid lateral motion during depressurisation.

The lateral strength of a rock berm is, however, limited and neither a rock berm nor a rock dumped trench is expected to provide sufficient lateral support to ensure that lateral motion does not take place during depressurisation.

When the pipe moves laterally the rock dump will fill the void behind the pipe. The lateral resistance of a rock filled trench will be nearly the same as virgin soil.

For a trenched and rock dumped pipe the only method that can be used to avoid upheaval buckling is to ensure that the lateral deflections have sufficiently small radius of curvature. The radius of curvature must be such that the force required to generate a lateral deflection is significantly smaller than the force required to generate an upheaval buckle. Use of large rock berm will help as the force required to generate upheaval buckles is thereby increased.

The same scenario as analysed for a trenched pipe with natural backfill (figure 5.1) has been analysed for a trenched and rock dumped pipe. The results are presented in figure 6.1, the soil properties used in the analysis are shown in figure 6.2.

By comparing figure 5.1 and figure 6.1 one will notice the following:

- The lateral motion during depressurisation will be similar in the two cases. This is a result of the lateral resistance which will be equal as soon as the pipe has penetrated the trench wall. This result is valid as long as the axial friction is the same.

- In both cases (natural backfill and rock dump backfill) upheaval deflection is avoided and lateral buckling is initiated.

- The force required to generate lateral deflections in a rock filled trench is 125 kN which is 80% larger than for a trench with natural backfill. The force is, however, still below the expected axial force required to generate an upheaval buckle (~ 200 kN).

- The resulting bending radius is 2.6 m in the rock berm case. Trenching with natural backfill gives a radius of 5.9 m. It should be noted that the feed-in length is conservative in both analyses, the bending radius will hence be larger in the actual application and overbending is not expected.

As shown lateral deflection may be obtained for trenched and rock dumped pipes. To ensure such deflection it is, however, required that the lateral radius of curvature is below a threshold value. Analyses with a straighter pipe during trenching have been performed. (10% of amplitude and same deflection length as used in the analyses reported in figures 5.1 and 6.1). A vertical trenching profile with 10 cm double amplitude and 50 m wave length has been included as a gentle vertical imperfection.

Two cases have been analysed; trenching with natural backfill and trenching with rock dumping. The pipe and soil parameters are as described in sections 4 and 5. The pipe is axially elongated with 0.2%. Table 6.1 presents the key results.

The flowline which is trenched with natural backfill will even in this relatively straight trenching configuration develop lateral deflections. The rock dumped pipe, on the other hand, will buckle vertically. The upheaval buckling amplitude is based on a rock cover with 13 kN/m^2 initial uplift resistance. The uplift resistance is reduced to the pipe weight when the pipe has moved 10 cm upwards. This effect combined with long feed-in length and the large axial elongation used in the analysis make the amplitude shown in figure 6.4 conservative.

Case	Trench with Natural Backfill		Trench and Rock dump	
Lateral deflection wave length during trenching (m)	50	50	50	50
Lateral deflection double amplitude during trenching (m)	10	1	10	1
Buckling mode	lateral	lateral	lateral	up-heaval
Buckling length (m)	20	13	10	15
Buckle initiation force (kN)	60	140	125	185
Force in buckle after elongation (kN)	40	50	40	30
Figure	6.1	6.3	6.2	6.4

SUMMARY OF 10" FLEXIBLE FLOWLINE BUCKLING CALCULATION

TABLE 6.1

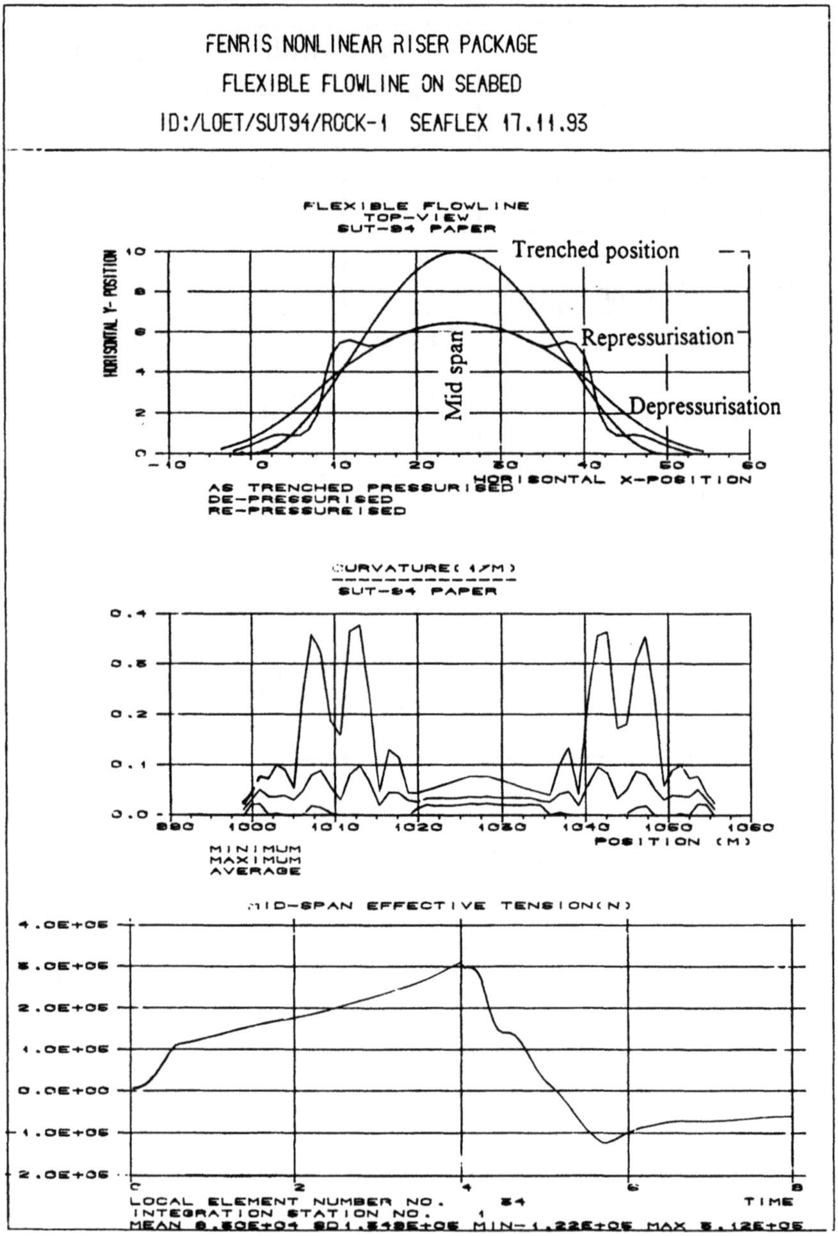

NON-LINEAR FENRIS ANALYSIS /2/ OF ROCK DUMPED FLEXIBLE PIPE SOIL MODELLED BY ELASTOPLASTIC PROPERTIES, REF. FIGURE 6.2

FIGURE 6.1

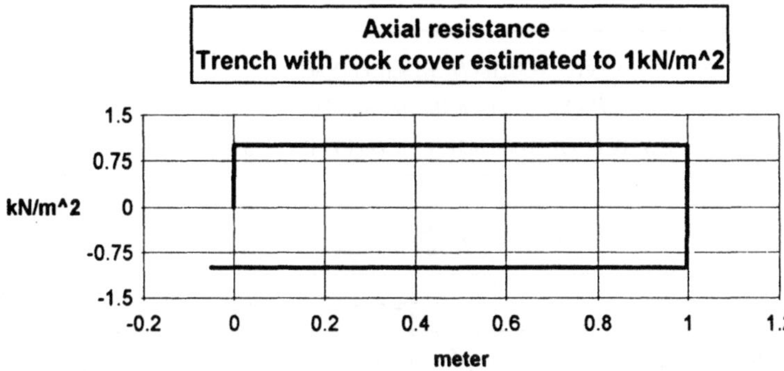

A) Axial Properties

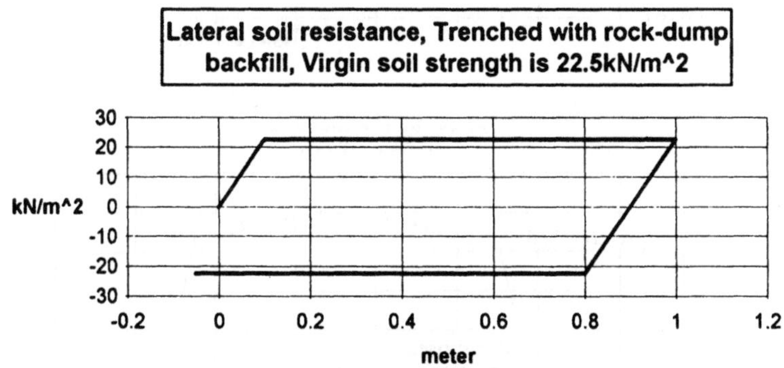

B) Lateral Properties

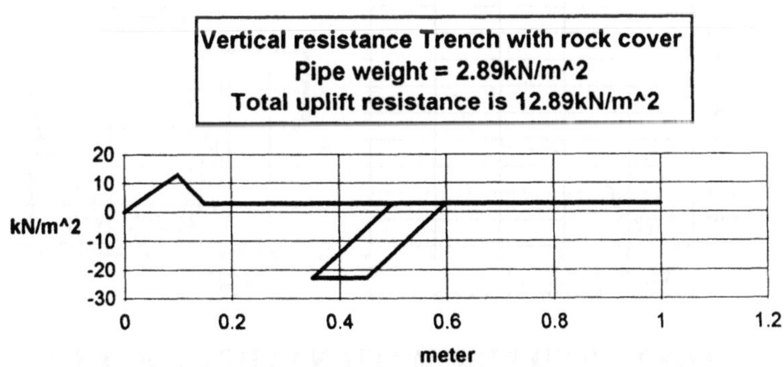

C) Vertical Properties

SOIL PROPERTIES ROCK DUMPED TRENCH MODELLED BY ELASTOPLASTIC SOIL SPRINGS

FIGURE 6.2

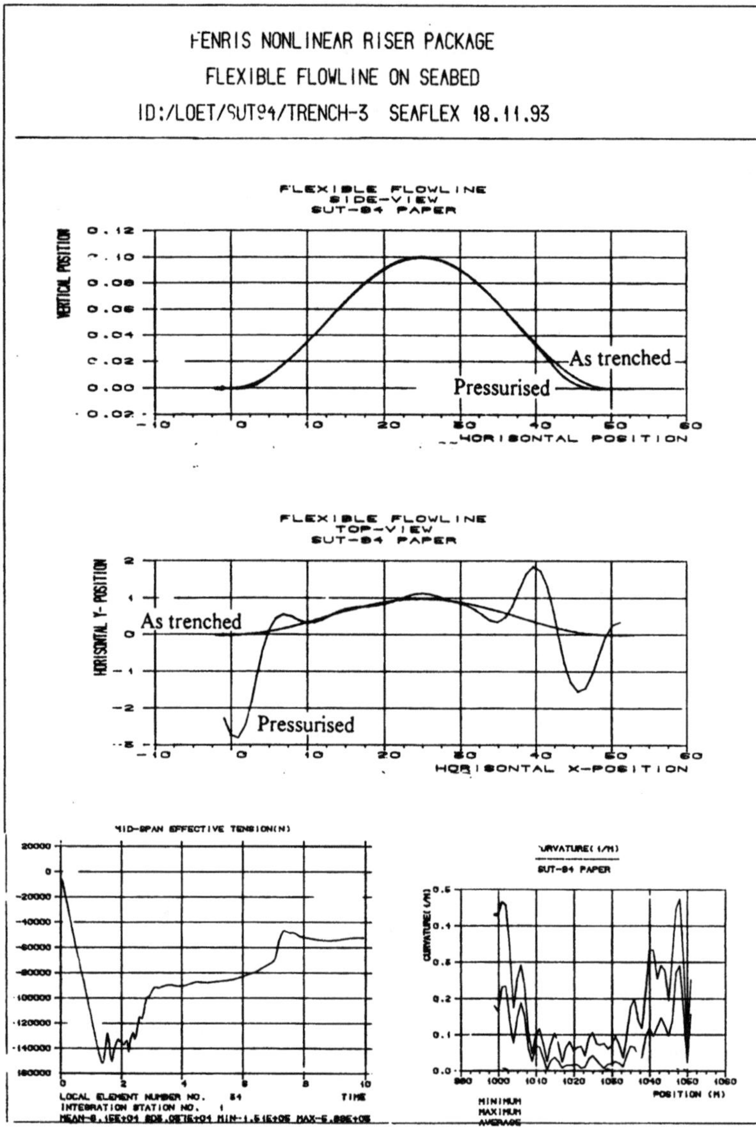

TRENCHED PIPE WITH NATURAL BACKFILL

NON-LINEAR FENRIS ANALYSIS /2/
50 m long initial lateral deflection with 1 m amplitude
50 m long initial vertical buckle with 0.1 m amplitude

FIGURE 6.3

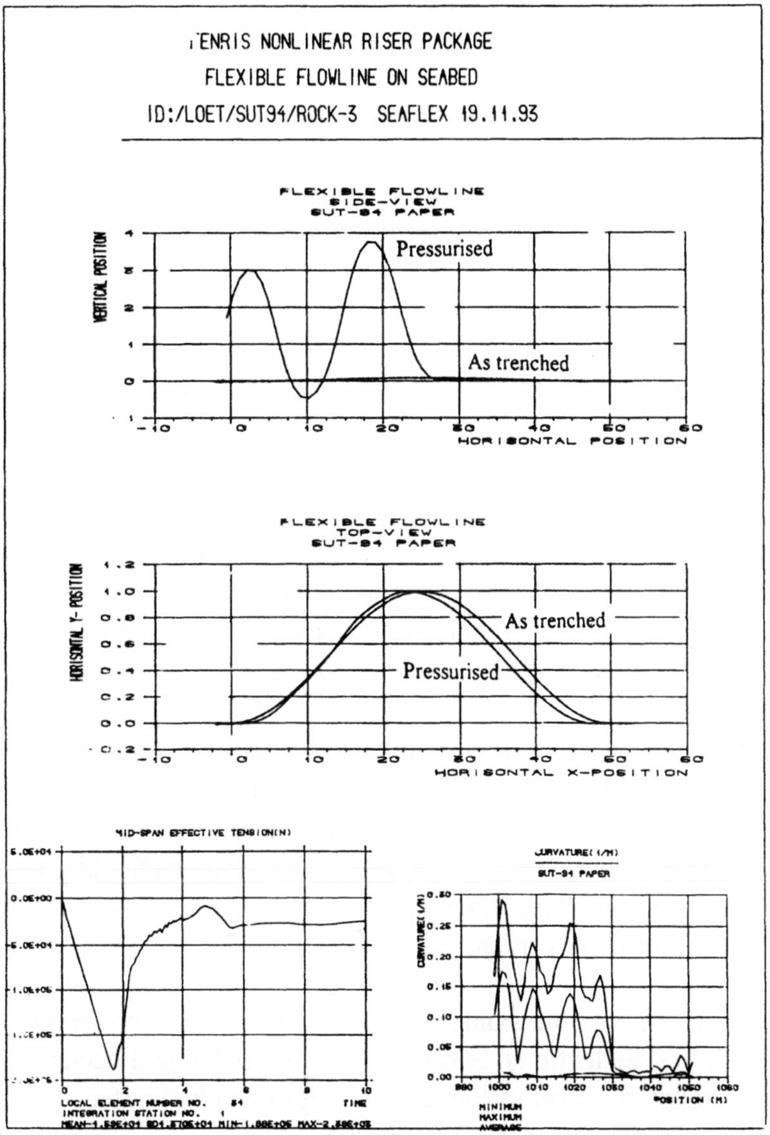

TRENCHED AND ROCK DUMPED PIPE

NON-LINEAR FENRIS ANALYSIS /2/
50 m long initial lateral deflection with 1 m amplitude
50 m long initial vertical buckle with 0.1 m amplitude

FIGURE 6.4

7. PROTECTION SOLUTION

7.1 Protection Solution for Gathering Lines and Long Flowlines

Based on buckling evaluation for the flexible lines the following was concluded:

- Trenching is recommended.
- The flexible lines should be trenched in a pressurised mode and the pressure should be as high as possible to ensure maximum axial elongation of the lines during production operations.
- The flexible lines should be laid in a snaked geometry providing natural lateral imperfections to obtain lateral deflection of the lines when being pressurised. When the lines are trenched or rock dumped in this condition the upheaval buckling potential is reduced.
- From an upheaval buckling point of view it is preferable to protect the flexible lines by trenching with natural soil backfill rather than by rock dumping. The pipe is to a certain extent free to expand and contract in terms of lateral movement in the trench and vertical buckling is avoided. The weight of the overburden reduces the potential to buckle vertically but allows lateral movement.
- The maximum possible axial force is limited to the force required to initiate lateral deflections. Should a vertical buckle still occur the amplitude will be small as the feed-in length will be limited.
- The trench should be relatively narrow for trawl board protection.
- The trench walls should be vertical to avoid semi vertical buckling.
- The trench depth should be relatively constant and local vertical imperfections should be limited.

The soil conditions in the Troll Field is in favour of trenching. The clay is soft but the shear strength and the plasticity ensures that trench walls will remain nearly vertical after trenching. Trenching by jetting and fluidizing in such soil conditions have proven to be very efficient and natural backfill has been achieved during trenching of the TOGI umbilicals bundle in the Troll area.

Numerous pockmarks (depressions) are present in the seafloor in the Troll area. These depressions are typically 50 - 100 m in diameter and 5 - 8 m deep. These depressions have been avoided by routing the lines in snaked corridors. In addition, detailed surveys of the chosen route corridors have reviled a number of small local depressions typically 1 m deep and 5 - 10 m wide. Local snaking of the lines with curve radius approximately 200 m is planned to avoid these depressions. Thus, the recommended snaked laying geometry is obtained in a natural way and the lateral deflections will be present already prior to pressurisation. Figure 7.1 shows typical pipe routing in areas with pock marks.

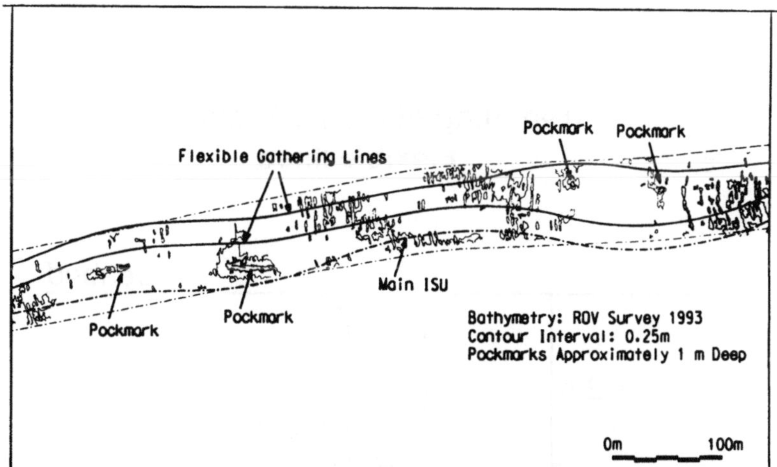

Snacked Routing of Flexible Lines to avoid Pockmarks

TYPICAL EXAMPLE OF PIPE ROUTING TO AVOID POCKMARKS

FIGURE 7.1

Trenching have been selected as the main protection solution based on the following conclusions:

- A narrow and deep trench with natural soil backfilling is considered as sufficient trawl board protection.
- Trenching with backfilling provides necessary additional thermal insulation.
- Trenching in pressurised snaked condition provides good protection against upheaval buckling.
- Trenching is cheap compared to rock dumping.
- Trenching is a quick and efficient protection method providing immediate protection after installation.

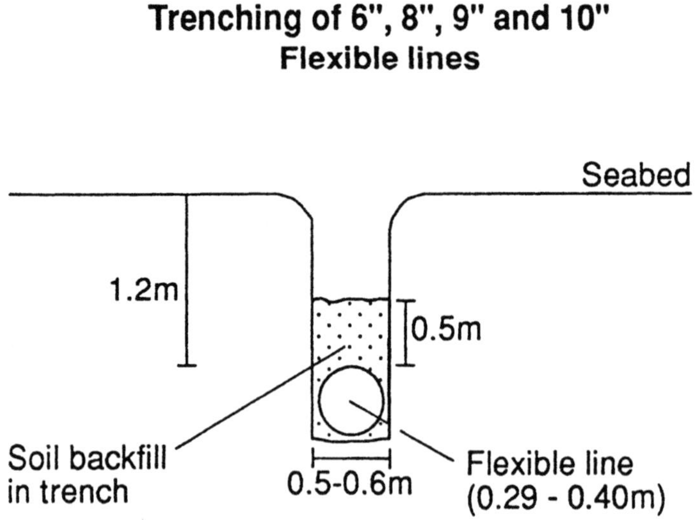

TRENCHING REQUIREMENTS

FIGURE 7.2

The trenching requirements are illustrated in figure 7.2. Maximum 0.3 m vertical imperfection heights are allowed in the trench. The lines will be trenched in their entire lengths between the riser base and manifold for the gathering lines and between the manifold and the remote satellite wells for the long flowlines.

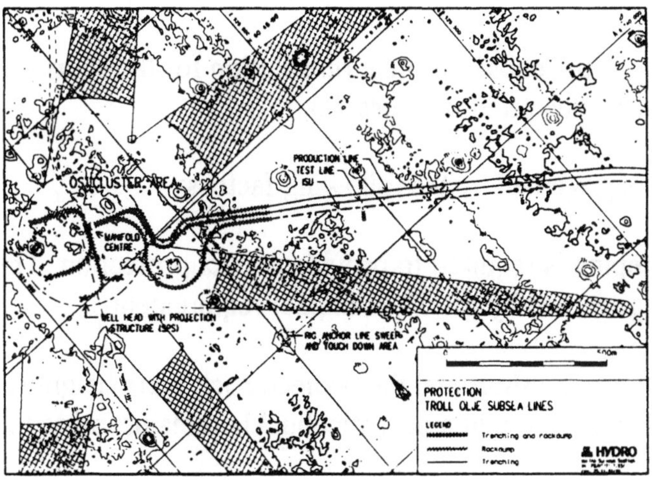

PROTECTION REQUIREMENTS

FIGURE 7.3

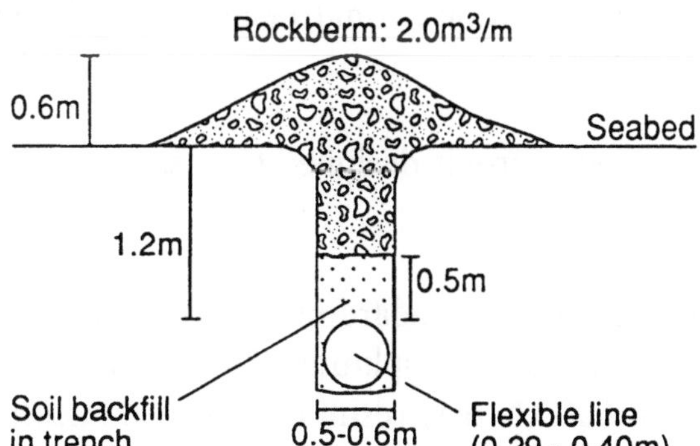

Combined trenching and rockberm protection

PROTECTION SOLUTION

FIGURE 7.4

The lines will be rock dumped on top of the trench to protect against dropped objects and anchor chain touch down in the vicinity of the cluster areas. The extent of these areas are illustrated in figure 7.3. The protection solution for the gathering lines and long flowlines are illustrated in figure 7.4.

7.2 Protection Solution for the Short Flowlines

Rock dumping is the chosen protection solution for the short flowlines. The requirements are illustrated in figure 7.5.

This solution is governed by the following factors:

- Rock dumping is required to provide impact protection. The rock berm height is governed by the 50 kJ impact requirement.

- The flowlines are short and the planned rock berm height is more than sufficient to ensure that a potential buckle remains inside the rock berm.

- Trenching is not practical and partly impossible due to the close spacing of the lines.

- Additional thermal insulation (from trenching backfilling) is not required for the short flowlines.

- The slope of the rock berm is designed to ensure stability of the berm in the soft clay.

An upheaval analysis based on standard methodology /1/ has been performed to check the maximum buckling amplitude. As shown in figure 7.6 a pressure induced axial force of 600 kN may result in upheaval buckling even with an overburden of 15 kN/m. With a large axial friction the buckling amplitude will be small. The friction resistance may, however, be reduced over time (creep). An axial friction of 0.1 will result in a buckling amplitude of 0.6 m which is acceptable in the proposed rock berm. As the feed-in length is limited for these pipes it is not expected that the pipe will creep out of the rock berm.

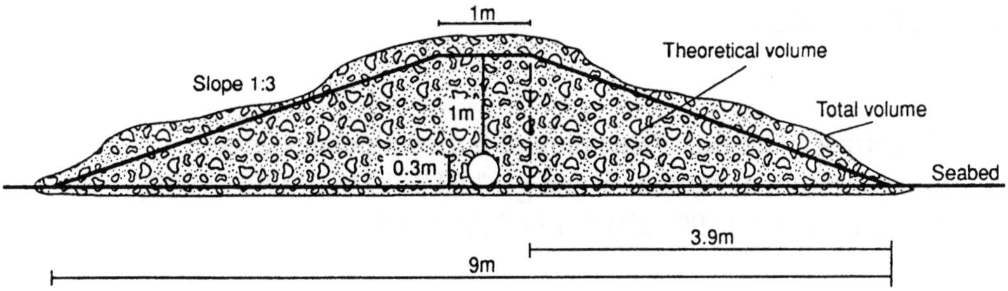

ROCK DUMPING REQUIREMENTS

FIGURE 7.5

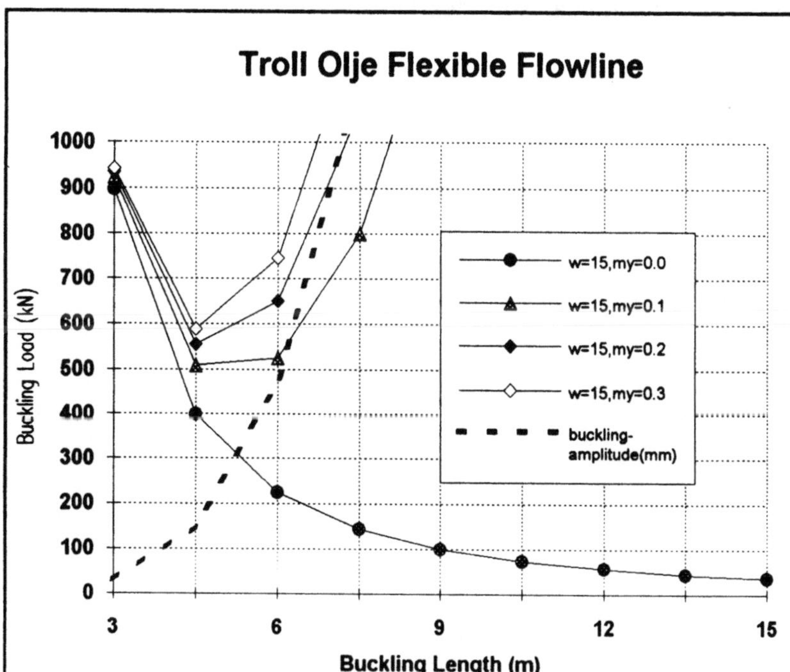

UPHEAVAL BUCKLING OF ROCK DUMPED SHORT FLOWLINES

FIGURE 7.6

8. REFERENCES

/1/ Guttormsen and Aas, "Buckling of buried pipelines under transportation of heated oil and gas",EUROMS-ISOPE-1990, Trondheim aug-1990, ISBN 0-9626104-4-5

/2/ Det norske Veritas, "Sesam User's Manual, FENRIS", 01-March-1993

/3/ Arild Bech and Nils Sødahl, "Structural Damping in Design Analysis of Flexible Risers", Marinflex 92, Flexible Pipes, Umbilicals, Marine Cables, BPP Technical Services 1992.

Session 3
Construction/Installation

REMOTE CONNECTION SYSTEMS FOR SUBSEA PIPELINE TIE-INS AND REPAIRS

DEREK CRUICKSHANK and PHIL MAXTED
Hydra-Tight Limited,
MORGRIP Product,
Howemoss Crescent,
Kirkhill Industrial Estate,
Dyce,
Aberdeen, AB2 0GN

ABSTRACT

A range of Mechanical Pipeline Couplings and Pressure Test Caps which are capable of connecting and sealing a wide range of steel pipework with diameters up to 48" and with internal pressures in excess of 10,000 psi has been developed and is now rapidly gaining market acceptance.

The Couplings have been designed to replace the more traditional methods of pipeline connection such as API/ANSI Flanges, clamp type connectors and even joining by welding.

The system is a bolted, gasketed Coupling that does not require any welding to provide a permanent connection that has been proven by Det Norske Veritas and Lloyds to be equally as strong as a butt welded connection.

The Coupling has been extensively third party tested and its performance has allowed engineers to specify its use on applications where mechanical connectors would not have previously been considered.

The MORGRIP coupling has been used on applications both Topside and for subsea pipeline tie ins and repairs.

The Coupling system is currently being adapted for operation by an R.O.V. which will eliminate the need for diver intervention and will offer a cost effective solution to the problem of deep water connections.

INTRODUCTION

The pressure on oil and gas operators to find and develop new resources is leading them to investigate methods of

Volume 33: Aspect '94, 113–128.

extracting oil from areas which up to now have been classed as inaccessible.

These fields lie in waters whose depth prohibit the use of installation and construction involving conventional saturation diving techniques.

The past few years has seen a significant increase in the systems and equipment available to perform many of the tasks required to develop the fields without the need for diver intervention.

For example, the Vetco Gray Remote Subsea Connection System named GRS with its single drive screw has been utilised on several flowline installations, However, this system like others that have been or are currently being developed rely on at least one of the mating connector components being pre-fitted / welded to the pipe on the lay barge prior to lowering Sub-Sea.

These systems allow for initial construction of deepwater installations but it often proves difficult or impossible to rectify damaged pipelines.

This paper offers a solution to the problem of making a connection Sub-Sea without diver intervention and where no surface pre-fitting of components is required.

i.e.

An ROV operated system can be deployed to cut away a section of pipe that has been damaged, remove any surface coatings, align the replacement pipework and finally connect by the use of a weldless coupling system. Trade-marked MORGRIP.

REPAIR SCENARIO'S

The development of the MORGRIP coupling towards a diverless connection system has been driven by STATOIL and the NORWEGIAN PETROLEUM DIRECTORATE who respectively control the majority of Norwegian pipelines and issue the directives covering their safe installation and operation.

The Oil Directorate concerned over the effects, both economic and environmental of a leak developing in a pipeline that is installed in waters whose depths are in excess of existing repair system deployment range are currently issuing a set of regulations.

These regulations will cover contingency plans that must be in place to cover for various scenarios relating to laying, operation and maintenance of the pipelines.

These directives will, firstly ensure that prior to pipe laying a system is in place to cover for the 'Wet Buckle' scenario.

The present methods are to either cut the line, retrieve it onto the barge , cut out the damage and to recommence welding.

This is obviously time consuming , the alternative is to cut the pipe and to commence laying next to the pipe subsea. This does not hold up the laying operation but it requires a welding spread to be mobilised to connect the two ends subsea.

In terms of deepwater both options cause problems as the pipe retrieval can lead to new buckles and it is currently not possible to weld beyond 360M.

In addition to pipe laying the Oil Directorate have stated that prior to a pipe line coming into operation a repair contingency must be in hand.

STATOIL, for their Haltenpipe projects, part of which are to be installed in areas whose depths are in excess of 500 m. had a repair plan such that new pipe would be stocked and in the event of a leak the pipe would be shutdown and a new line laid.

SUBSEA CONNECTION SYSTEMS

In order to select the connection system best suited to the repair of a damaged pipeline the current methods of pipeline connection must be evaluated and their ROV adaptability investigated.

these being ;

SPLIT TYPE CLAMPS.

Primarily used for repairs to dents and pin hole leaks.

HYPERBARIC WELDING.

Favoured method of connection subsea for tie - ins and repairs.

Problems exist with very deep water due to technology and cost.

MECHANICAL CONNECTORS.

(including standard flanges).

These can be divided into connectors such as flanges, Graylocs etc which require to be pre welded to the pipe.

The problem is then one of installing and tensioning bolts subsea.

The second category is that of the truly mechanical connector, one which has the ability to mechanically lock and seal.

The selection of the repair method depends upon the type of damage and the pipelines location, its contents and the speed of response required.

Split clamp type repair systems are positioned around the pipe to be repaired. The unit`s bolts are tensioned and the clamp is then injected with some form of grout.

This repair is suitable for strengthening and small pin hole type leaks.

The mechanical connector has been used to replace hyperbaric welding for less critical applications due to the reduction in overall repair costs.

The existing mechanical connectors are only classified as temporary repairs due to their design and lack of approvals and compliance to the industry design standards.

In addition, their use has been limited due to the lack of operator confidence in the products in terms of its long term reliability.

This lack of confidence has partially arisen due to problems associated with the older type units.

This has led to Hyperbaric welding becoming the preferred method of repair and construction especially as the weld repairs the pipe back to its original strength.

The problem that exists with hyperbaric welding is its current inability to be used economically in water greater than 350 metres deep.

The mechanical connector for deep water repairs appears to be the only viable option, especially now that the confidence barrier has been overcome by the introduction of the MORGRIP mechanical connection system which is approved for use by many of the certified authorities and is classified as a permanent connection equal in strength to a butt welded joint.

THE MORGRIP WELDLESS CONNECTION SYSTEM

DESIGN

One major obstacle in the path of using a product such as the MORGRIP coupling is the lack of laid down specifications in the standards that cover the industry. This in turn prevents designers from using the product as they have no real means by which to judge its design.

Specifications that cover the design of new pipe work systems in terms of material selections, bolt and gasket loadings, testing etc. are ambiguous and if applied to existing technology, i.e. ANSI flanges then they would not be considered safe to use due to their inability to comply with the full requirements, for example proof testing.

The standard bolted connection methods such as the ANSI flange and clamp type connectors can leak if incorrectly installed or in situations where unforeseen loads are applied. i.e pressure shocks, for this reason in many areas on critical pipelines the bolted connection is not preferred.

The uncertainty about their possible performance has led the industry away from using them on critical pipelines subsea, towards hyperbaric welding, especially in Norway.

It appears from market research that the decision whether to weld or to install a flanged system is inconsistent and is affected by many factors.

The cost of hyperbaric welding is ever increasing and when coupled with the current depth limitation of welding it has led to operators to look again at bolted connections for subsea tie-ins.

Investment is being made into methods of remotely tightening bolted connections but these systems cannot be utilised for repair applications.

The MORGRIP coupling has evolved to replace the welded connection in both topside and subsea applications. The product is gaining market acceptance especially for installation driverless.

Prior to describing the couplings ROV development the next section outlines the products basic design.

The MORGRIP has been developed by carefully controlled research development and testing programme which has been overseen by both Lloyds and D.N.V.

The past three years have seen the product gain an extensive track record on all types of process pipework, both topside and subsea.

THE PRINCIPLES OF OPERATION

(refer to Figs 1, 2 & 3).

The coupling has a modular design. It comprises a centre section housing, one or a series of griping segments on either side of the centre section. The number of gripping segment pairs applied depends on the pressure rating for the coupling. Each gripping segment houses a series of spring loaded ball bearings positioned circumferentially around the inside of the segment bodies.

To connect two lengths of pipe, one grease plate and one or more gripping segments are slid over the end of each pipe. The centre section housing is then positioned over the pipe joint between the gripping segments. All segments and centre housing are then pulled together by studbolts located within the coupling. When the bolt load is applied, the ball bearings roll and swage into the external surface of the pipe thus gripping it.

The ball bearings being independent of each other are capable of compensating for oval pipe and any surface imperfections.

The indentation of the ball bearings is limited to a maximum depth of 12.5% of the pipe wall thickness or to a maximum of 2 mm.

The indentations are formed by a ball bearing and as such do not act as stress raisers in the pipe as proved by the fatigue testing undertaken by TWI (5).

SEALING SYSTEM

The MORGRIP incorporates two sealing barriers located onto the periphery of each of the two pipes to be connected. These radial seals are located within separate seal housings which are independent of the gripping system.

The radial seals are compressed and seal on the pipe`s outer surface when the bolts are tensioned.

Fig 1 also illustrates the radial sealing system which is a soft seal with metal anti extrusion rings.

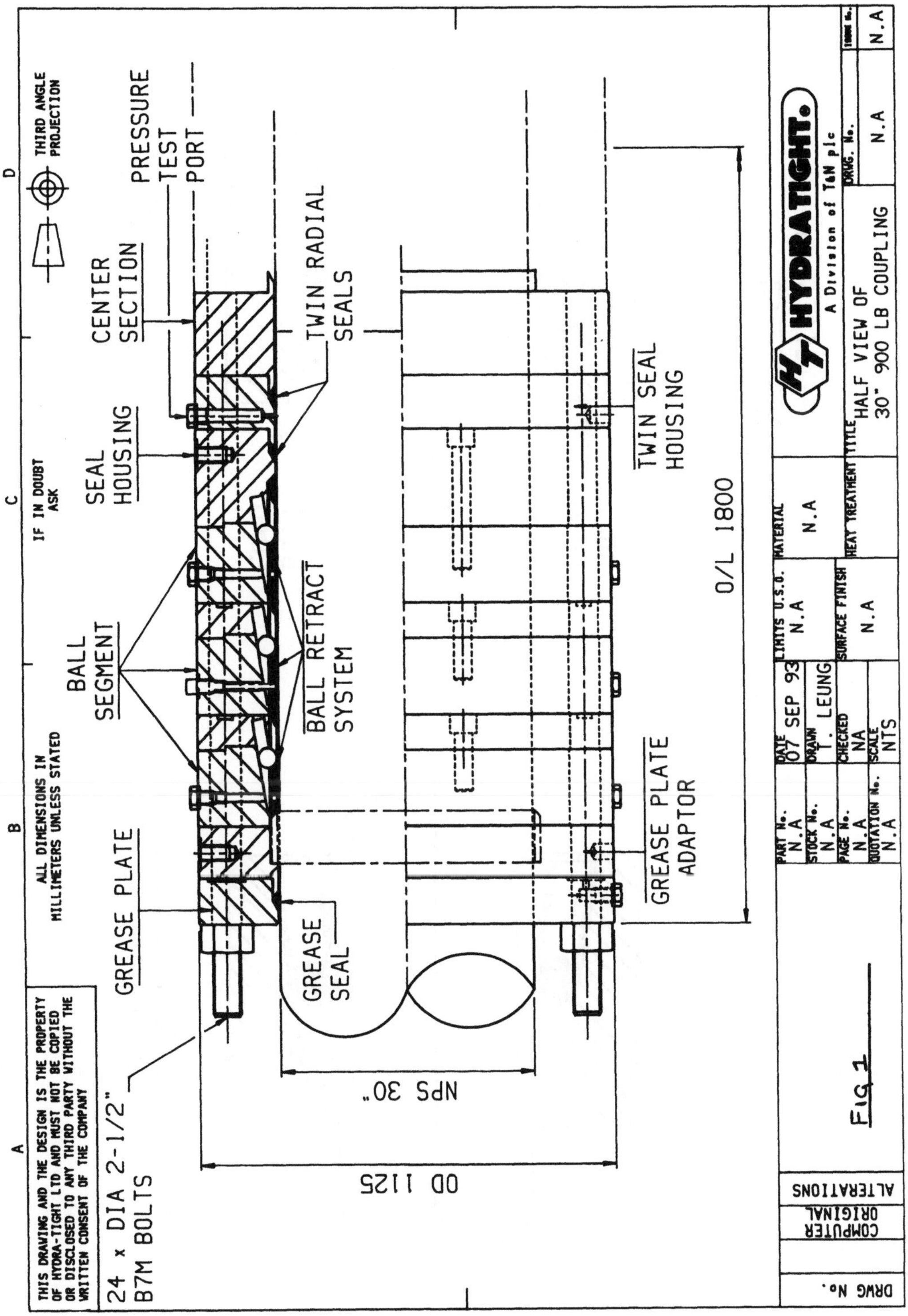
THIS DRAWING AND THE DESIGN IS THE PROPERTY OF HYDRA-TIGHT LTD AND MUST NOT BE COPIED OR DISCLOSED TO ANY THIRD PARTY WITHOUT THE WRITTEN CONSENT OF THE COMPANY
ALL DIMENSIONS IN MILLIMETERS UNLESS STATED
IF IN DOUBT ASK
THIRD ANGLE PROJECTION
24 x DIA 2-1/2" B7M BOLTS
GREASE PLATE
BALL SEGMENT
SEAL HOUSING
CENTER SECTION
PRESSURE TEST PORT
TWIN RADIAL SEALS
BALL RETRACT SYSTEM
GREASE SEAL
NPS 30"
OD 1125
O/L 1800
TWIN SEAL HOUSING
GREASE PLATE ADAPTOR
HYDRATIGHT
A Division of T&N plc
PART No. N.A
STOCK No. N.A
PAGE No. N.A
QUOTATION No. N.A
DATE 07 SEP 93
DRAWN T. LEUNG
CHECKED NA
SCALE NTS
LIMITS U.S.D. N.A
SURFACE FINISH N.A
MATERIAL N.A
HEAT TREATMENT
TITLE HALF VIEW OF 30" 900 LB COUPLING
DRWG. No. N.A
N.A
ALTERATIONS
COMPUTER ORIGINAL
DRWG No.
FIG 1

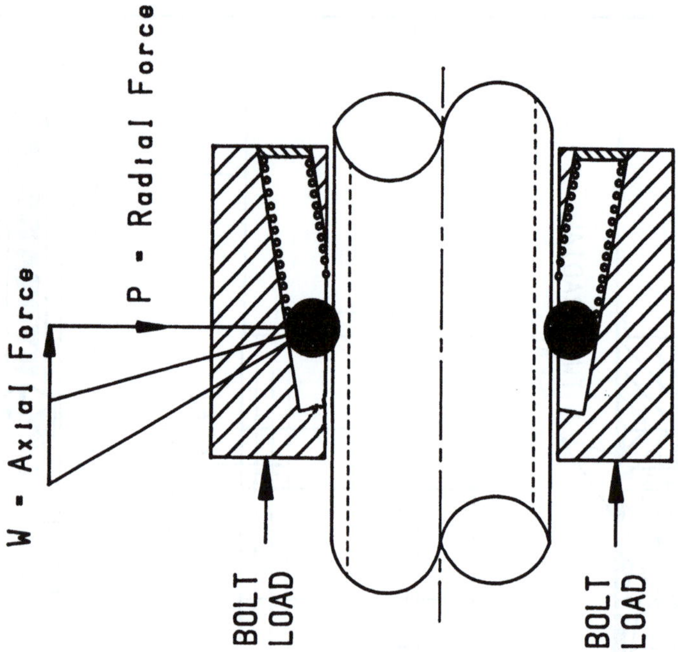

Fig.3. After Application Off Bolt Load.

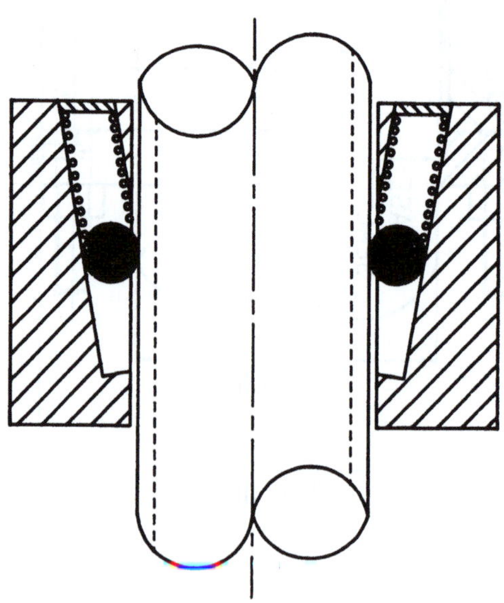

Fig.2. Before Application Off Bolt Load.

fig 4

Schematic Showing Possible Installation of a Hydraulically Operated MORGRIP Coupling

Deep water ruptured pipe. (600 Meters plus)

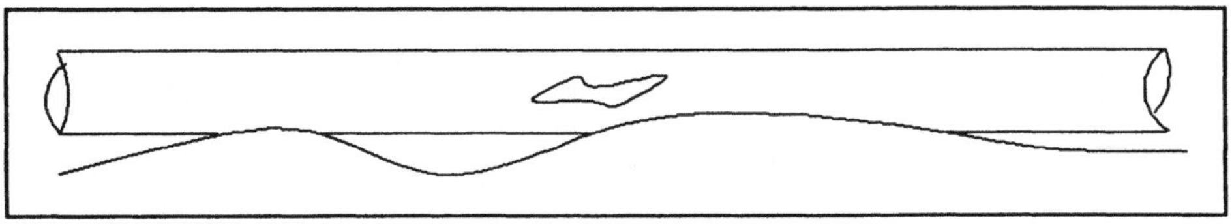

Pipe lifted off seabed and placed on two frames one either side of the rupture.

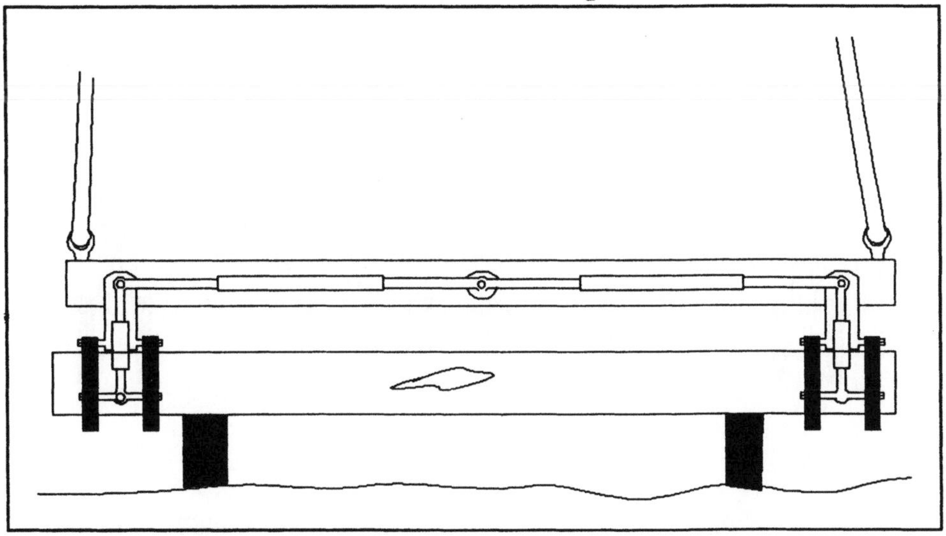

Pipe cutting equipment installed and the ruptured part of the pipe removed.

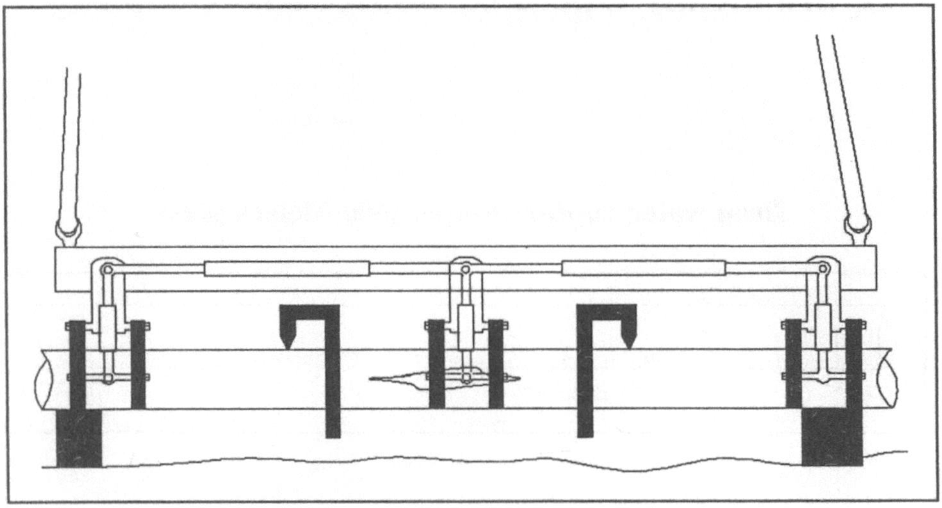

The pup piece is lowered into position with the two Hydraulic MORGRIP Couplings on either end.

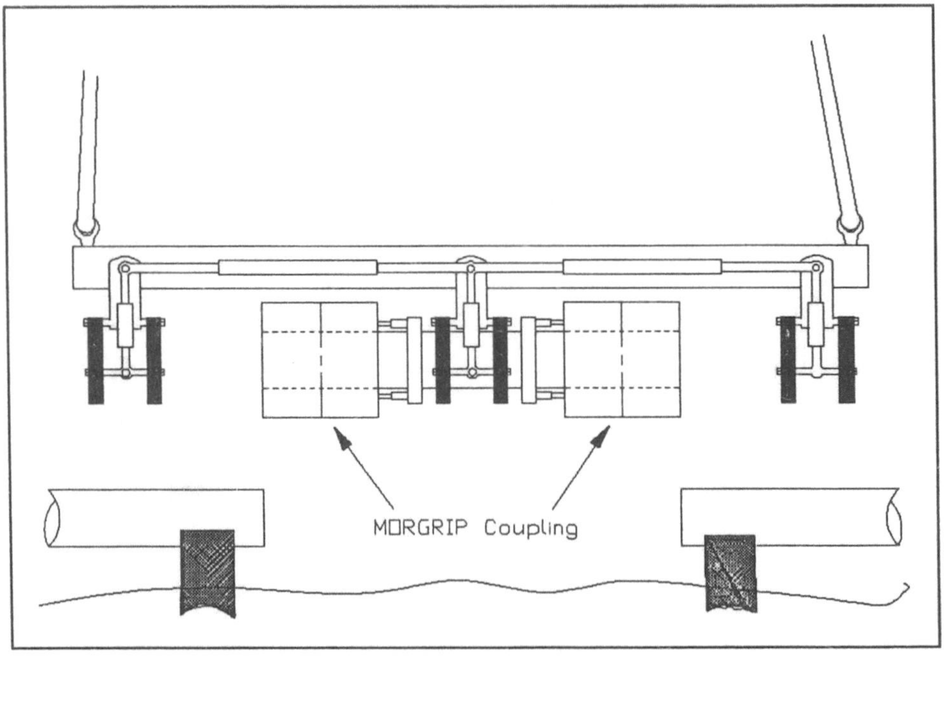

Finally the ROV attaches the umbilical and the MORGRIP Couplings are operated.

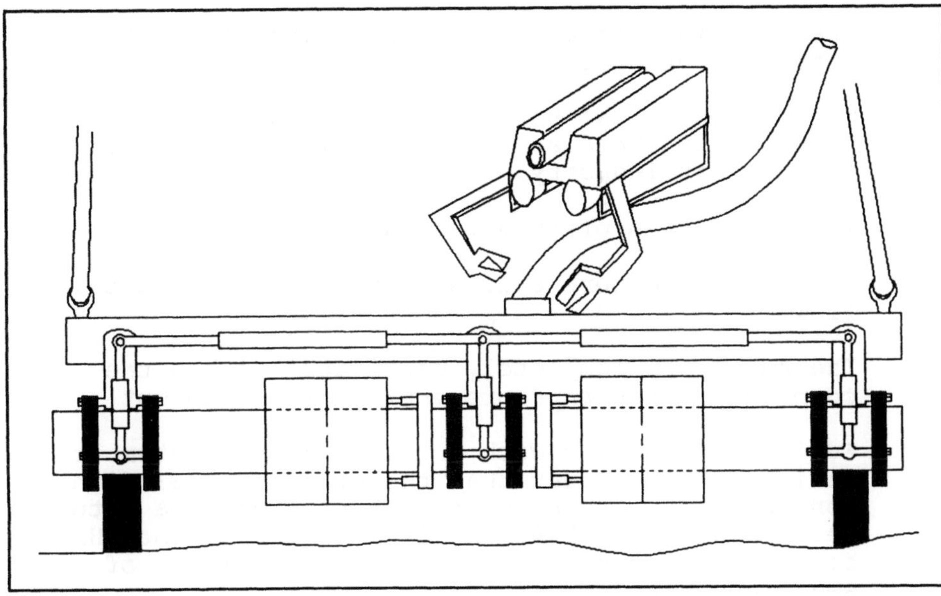

The initial tensioning compresses the gaskets to initialise the seal, the pressure energising effect compensates for possible pressure surges and overloading.

The advantages of a twin sealing system are as follows:

* The coupling can be externally tested for joint integrity prior to the internal pressure test.

* The external test port can be used as a monitoring point as is the case with the twin gaskets used in the nuclear industry.

* The twin seals offer a back up for safety.

* A smooth pipe bore exists thus preventing turbulent flow.

* The seals are completely captivated within the coupling`s body thus preventing seal blow out.

* Fugitive emissions with a twin sealing system are believed to be lower than the standard ANSI flanged connections. (this is to be confirmed by a future test program. HYDRATIGHT is a member of the BHR consortium researching into the performance of static seals and gland packings).

The coupling when tensioned becomes effectively `solid`, as such the unit is forgiving of over loading of the bolting as no flange rotation occurs.

The coupling behaves in terms of bolt loading as per the controlled compression bolted gasketed joint described by Nau Sept 1990 (1), with the exception that the radial sealing system compensates for gasket relaxation by being pressure energised.

The MORGRIP principle overcomes many of the inherent weakness of other methods of connection.

Advantages:

* Long bolt grip lengths to minimise relaxation losses (refer to Fig 4).

* Metal to metal flange contact to eliminate flange rotation and to offer a compression stop for the radial sealing system.

* Twin sealing system to allow for external pressure testing, joint monitoring and back up.

* Bolt stresses higher than that required to seal to compensate for external load considerations.

* A total connection that is considered to be as strong as an equivalent welded connection.

* A fire safe connection when exposed to a hydrocarbon fire.

* Weldless installation.

* Capacity to compensate for external load conditions including fatigue.

ROV DEVELOPMENT

The MORGRIPS design lends itself to installation by an ROV, but its ability to solve the problem is based upon the whole system.

The system includes the pipe cutting and the weight coat removal equipment, the alignment frames and the hydraulic supply.

As part of the development project the requirements of the coupling are being evaluated with a view to modification of the equipment already available or a complete design and development of new handling frames etc.

It is envisaged that the prototype coupling will be ready for trials in April 1994 with the first production unit being available in July.

PRINCIPLE

The standard MORGRIP design as detailed above relies on stud bolts being tensioned to actuate the gripping and sealing mechanism.

The stud bolts are simply removed from the coupling and an internal hydraulic actuation system is installed with a mechanical backup.

This actuation system allows the coupling to be installed remotely when coupled with the following features.

Ball Bearing Retraction System

The coupling is designed to be fully re-useable which is achieved by the use of a ball bearing retraction mechanism.

This takes the form of an internal cage which when moved allows the ball bearings to retract into their housings.

When the balls are retracted, the bore of the coupling is clear and smooth whose diameter is greater than the maximum allowable pipe diameter to which it is to be fitted.

The coupling is installed as one solid unit even though its design allows it to be installed in segments for difficult access situations.

The solid coupling is simply slid on to one of the ends of the pipes to be connected, or it is located on to the replacement spool piece prior to lowering subsea.

Once the two pipe ends have been drawn together the coupling is centralised over the pipe ends by the use of suitable hydraulic positioning tools pressurising the coupling, realising the inner ball cage allowing the bearings to contact the pipe and allowing them to swage into the pipe`s surface and compressing the seals.

Once installed the hydraulic pressure is released after the mechanical lock has been actuated.

The coupling`s removal is simply the reversal of the pressurisation process.

Installation

The MORGRIP coupling is obviously a part of the overall deep water repair equipment. A diverless MORGRIP has initially been designed to be interfaced with Statoil`s existing pipeline repair system handling frames and pipe preparation equipment. However a dedicated installation frame will be manufactured to actually install and position the coupling on to the subsea pipework.

A typical installation would consist of the ROV operated cutting machine being deployed to cut away the damaged section of pipe.

The existing equipment deployed for Statoil`s PRS is based upon hyperbaric welding up to a depth of 350 m, is also capable of removing the pipe`s outer protective coating back to bare metal which is the condition that the MORGRIP requires.

Accuracy of the pipe cut is not calculated as previously described in The MORGRIP Weldless Radial Sealing System.

The pipe handling frames are capable of letting the pipe attach themselves to it backing up with the help of H frames and aligning the replacement spool piece with the parent pipe, the replacement spool having previously been cut to length and deployed subsea.

Depending upon the location of the repair, the two off MORGRIP couplings can be pre-fitted on to the replacement spool prior to lowering subsea and then aligned so they can be assembled on to the parent pipe prior to the spool piece being inserted.

The MORGRIP is installed as a complete unit fitted with its seals and is designed to suit the standard tolerances of the pipeline.

Once in position, the coupling is hydraulically actuated either by a surface run umbilical or by a stabbing from the work horse ROV.

The ball retraction system and the gripping system actuation are all hydraulic functions which have a mechanical locking system which actuates once the coupling has been energised.

The hydraulic pressure is then removed and the coupling can be checked for joint integrity through the test ports, again performed by the ROV. In the unlikely event of incorrect sealing the coupling can be removed simply by pressurising in the reverse port to release the system, the gasket changed either subsea or by recovery to the surface and the operation repeated.

The second phase of the project will concentrate on pipe that has moved during installation and therefore requires misalignment of the position.

It is hoped that this phase of the work will be completed by the year end.

THE FUTURE

The system once fully developed can be used to perform not only deep water repairs but for initial construction and for shallower water applications.
Hydra-Tight intend to have a full sized prototype ready shortly to undertake diving trials.

The above system is only part of the overall installation equipment required, such as pipe cutting, alignment etc.

This ancillary equipment is currently available and we feel easily adaptable to the needs of the coupling.

REFERENCES

1. B.S. Nau Controlled Compression Bolted Gasketed Joints September 1990

2. D.N.V. technical report No. 91-3639 Burst Testing of Morgrip Flange Couplings

3. API Specification 6FB Fire testing for end connections

4 ASME Divison VIII Pressure vessel design codes

5. The Welding Institute fatigue testing of Morgrip Flange Coupling R. Phaal June 1991

6. DNV Technical Report No Fire Testing of Morgrip Flange Couplings

7. Report on Internal Pressure Proof Test of the Morgrip Flange Coupling

8. British Gas PLC report on The Morgrip Flange Coupling

9. An Introduction to the design and behavious of bolted joints John Pickford Published by Dekker

10. Notes on the load carryhing characteristics of Pre-tensioned bolts - Tensioned joints P.J. Gill

11. API Specification for line pipe 5L

12. Fatigue design rules for welded steel joints T.R. Gurney Welding Institute Research Bulletin Vol 17 May 1976

13. BS5400 Part 10 1980 Steel, concrete and composite bridges part 10. Code of practice for fatigue British Standards Institute London 1980

ROV ASSISTED GEOTECHNICAL INVESTIGATION OF TRENCH BACKFILL MATERIAL AIDS DESIGN OF THE TORDIS TO GULLFAKS FLOWLINES

P.T. POWER and
R.A. HAWKINS
Fugro Limited,
18 Frogmore Road,
Hemel Hempstead,
Herts. HP3 9RT

H.P. CHRISTOPHERSEN
Saga Petroleum a.s.,
P.O. Box 490,
N-1301 Sandvika,
Norway

I. McKENZIE
Fugro-UDI Limited
Denmore Road,
Bridge of Don
Aberdeen AB23 8JW

Abstract. During 1992 an innovative geotechnical investigation was performed in order to assist the design of the Tordis to Gullfaks 'C' flowlines. Data was required for the analysis of upheaval buckling resistance within a back filled trench and to assess the effectiveness of the backfill material as a thermal insulator,
In order to make reliable predictions a comparable trenched section of the previously installed Snorre to Stratfjord pipeline was investigated. Samples of the backfill were taken by means of a boxcorer and laboratory tests were performed on the vessel and onshore. In situ testing for thermal conductivity and shear strength were performed by means of a Heat Flow Probe and an Insitu Shear Vane respectively; both of which were deployed from a Super Scorpio ROV.
The results of the investigation confirmed that the natural trench backfill would provide sufficient upheaval resistance and thermal insulation without resorting to gravel dumping.

1. INTRODUCTION

The Tordis subsea development will comprise a series of satellite wells linked to a Central Manifold. Export and water injection flowlines and an umbilical bundle will connect the manifold to the Gullfaks 'C' structure, as shown on Figure 1.

The 'Base Case' concept for the flowlines installation involved them being laid on undisturbed seabed and covered by a gravel berm, as shown in Figure 2. The function of the gravel cover was threefold; providing:

(a) Protection from external loads such as trawl boards.

(b) Resistance to upheaval buckling.

(c) Limited thermal insulation

A 'Revised Concept' proposed by Stena Offshore - the EPIC Contractor for the Tordis Flowlines - involved the trenching of the flowlines and backfilling with natural seabed material.

Volume 33: Aspect '94, 129–141.

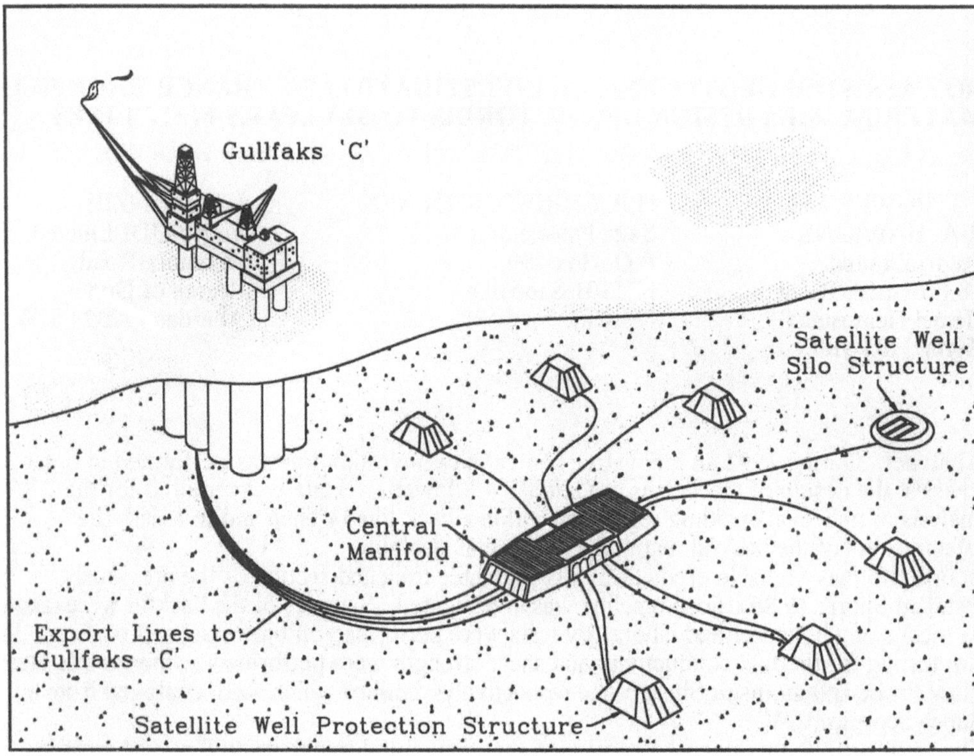

Fig. 1. Tordis Subsea Development

Verification of the acceptability of this potentially less expensive alternative was required to ensure that the natural backfill would be able to counteract upheaval buckling loads and give adequate thermal insulation. Soils data were available along the proposed route and hence the strength and density of the undisturbed soil were known. What was not known was how these parameters would change when the soil was disturbed and redeposited and what the thermal conductivity of this reconstituted material would be.

Since the Snorre to Statfjord pipeline had been trenched and backfilled in a similar fashion and in similar soils it was considered that it might present the opportunity to investigate the impact of the trenching and backfilling operation on the physical characteristics of the seabed soil. A brief desk study was undertaken to cross check soil types and a section of the Snorre to Statfjord pipeline in similar water depths to the Tordis flowlines (i.e. around 200 metres) was identified, see Figure 3. A decision was taken to mount a soils investigation that would involve insitu testing and sampling in the trench combined with a programme of laboratory tests on the soil samples.

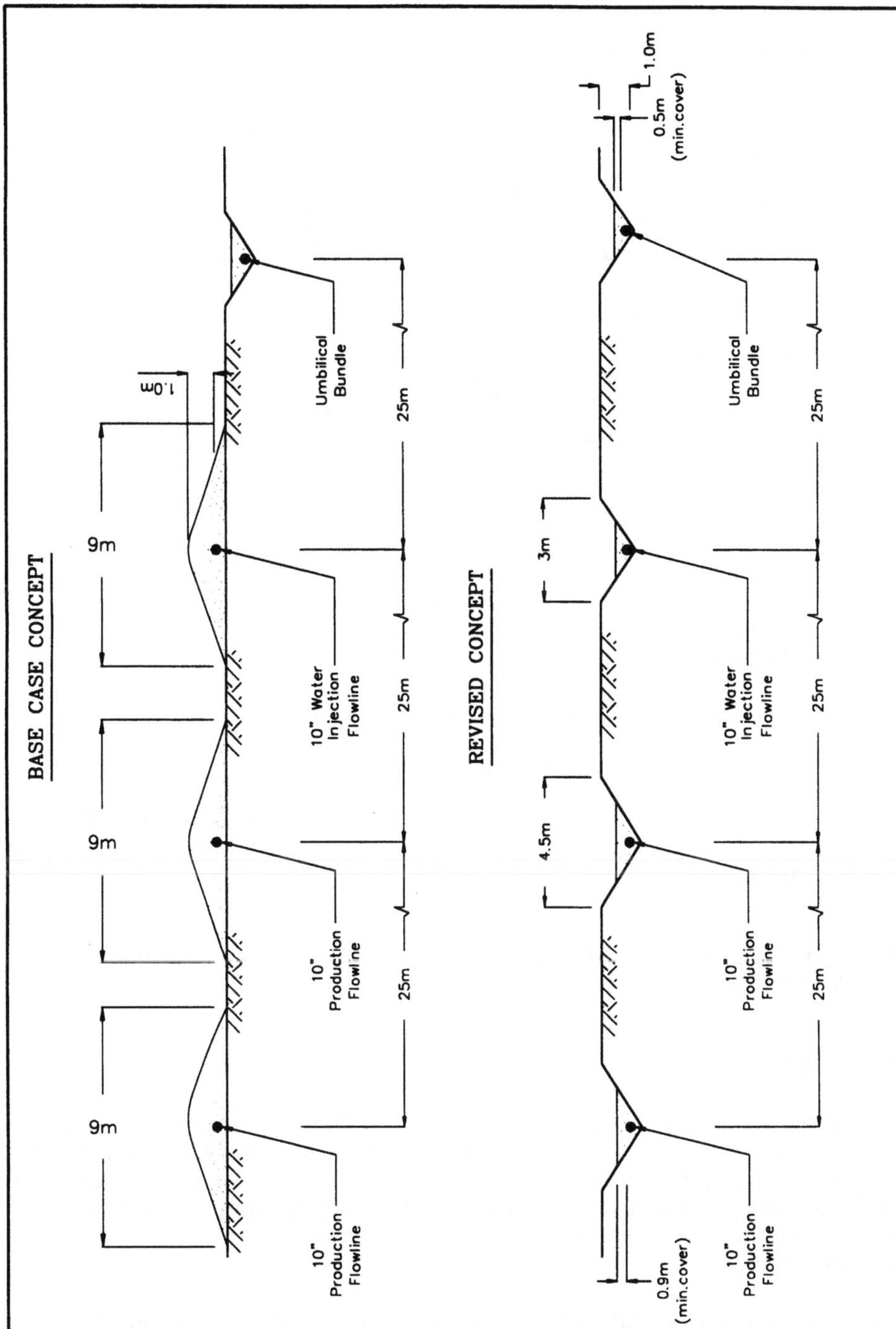

Fig. 2. Tordis Flowlines and Umbilical Installation Concepts

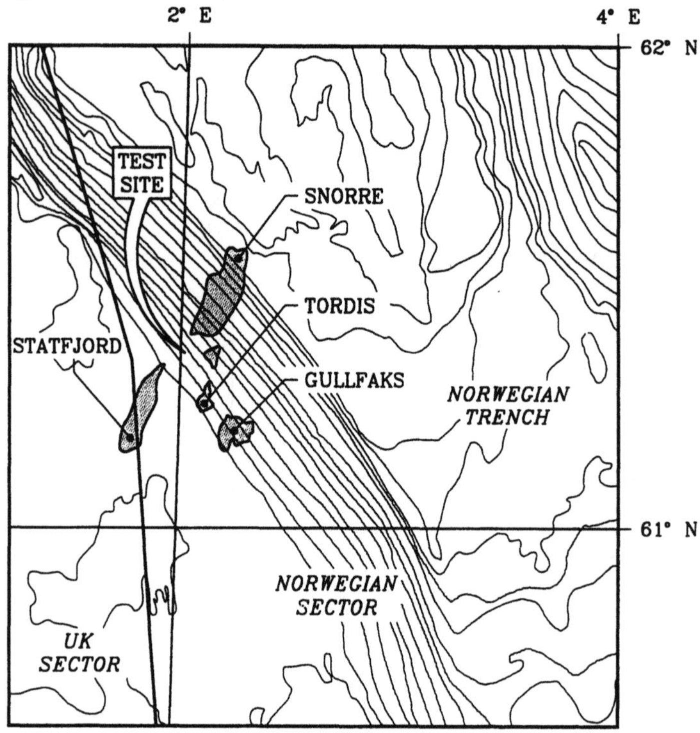

Fig. 3. General Location Plan

The undisturbed soil conditions along the Tordis route comprise a surface veneer (0.2m - 0.6m thick) of MEDIUM DENSE to DENSE fine to medium SAND with shells, overlying VERY SOFT CLAY containing thin sandy layers and having an undrained shear strength in the range 2kPa to 10kPa.

The soils investigation work was added to the scope of a planned pre-installation route survey for the Tordis flowlines.

2. SITE WORK

Marconi UDI (now Fugro-UDI) was contracted by Stena Offshore to undertake the pre-installation survey of the proposed 11.25km pipeline and umbilical route between the Gullfaks 'C' platform and the Tordis Manifold location. The survey, which was performed from the dynamically positioned vessel M/V Northern Surveyor, incorporated sidescan sonar and sub-bottom profiling and ROV visual and bathymetric surveys of the central proposed flowline route.

Fugro were sub-contracted by Marconi-UDI to provide a soil sampling and in-situ testing spread, comprising the following:

- A Seabed Box Corer
- An In-situ Vane Shear apparatus
- An in-situ Heat Flow (Thermal Conductivity) Probe
- An offshore soils laboratory
- Geotechnical personnel

Sampling and insitu testing was undertaken at three selected locations on the Snorre to Statfjord pipeline route as shown in Figure 4.

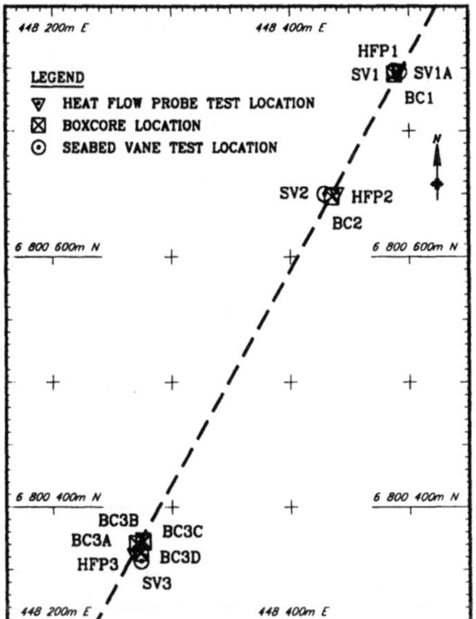

Fig. 4. Detailed Location Plan

2.1 Box Coring

The Boxcorer, a 'Reineck' type (see Figure 5) was deployed over the vessel's stern by means of a small winch and 'T' bar. Positioning the corer over the trench was achieved with the aid of an acoustic transponder on the corer frame and ROV observation. On retrieval, each sample was extruded from the core box and sections were tested for moisture content, bulk and dry density and shear strength, the remainder being subsampled and resealed for later transport to an onshore laboratory. Strength tests offshore were performed with a 'fall cone' apparatus and 'Torvane' hand held shear vane.

2.2 Insitu Vane Shear Tests

The Insitu Vane Equipment shown in Figure 7, and described by Geise et al. (1988), was deployed via the UDI "Super Scorpio" ROV, Figure 6. The vane was held in the manipulator of the ROV as it scanned the trench for suitable test locations. At the selected positions the vane was pushed into the soil to a penetration of around 0.5m. The testing

Fig. 6. Super Scorpio ROV

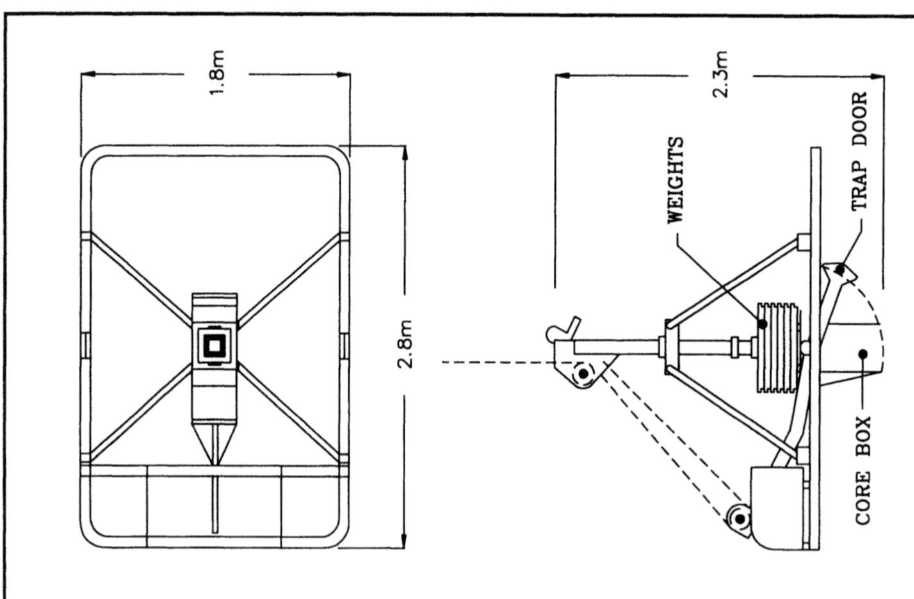

Fig. 5. Box Corer

sequence was initiated by the operator at surface and an integral electric motor in the equipment shaft rotated the vane blades at a constant speed, while instruments measured the rotation and the torque induced by the soil resistance. The test was continued until the soil sheared or 'failed' or the torque levelled off at a residual value.

The shear strength of the soil is derived from the relationship between the maximum torque applied and the surface area of soil that is sheared. Since the dimensions of the vane blades are known the sheared surface area can easily be calculated. A typical result from the investigation is shown in Figure 8.

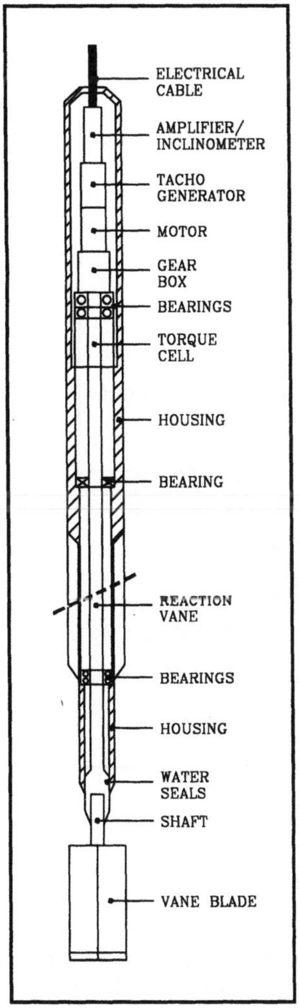

Fig. 7. In Situ Vane

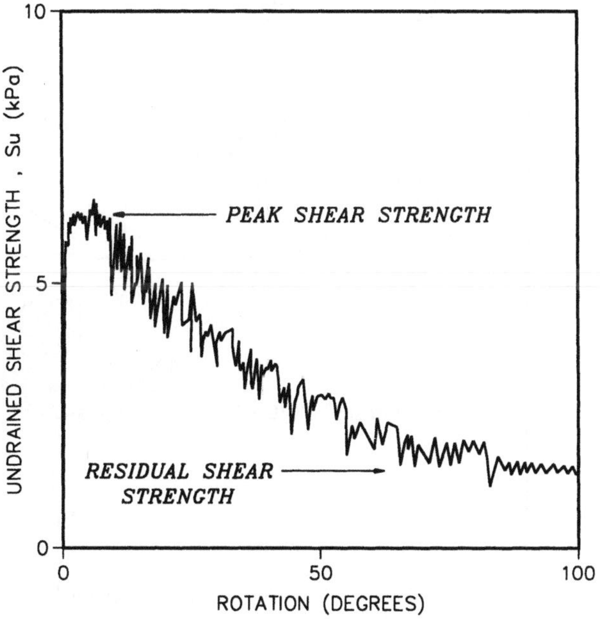

Fig. 8. In Situ Vane Test Result

2.3 Thermal Conductivity Tests

The Fugro Heat-Flow Probe, as described by Zuidberg et al. (1987) and shown schematically in Figure 10 was deployed in the same manner as the insitu shear vane i.e. via the manipulator of the ROV; see Figure 9.

At the chosen location the probe was inserted approximately 0.25m and the thermal conductivity test performed in two stages. The first involved the stabilisation of the probe temperature so that it was in equilibrium with that of the seabed. This was monitored via thermistors in the probe tip and usually took about 10 minutes. The second involved heating the surrounding soil by means of the internal heating element and measuring the temperature rise with time, usually for a period of about 10 minutes. A typical test result plot is shown in Figure 11. The thermal conductivity of the soil is calculated from the gradient of the curve where steady cylindrical heat flow has been achieved, using a formula that incorporates the known thermal resistance of the probe.

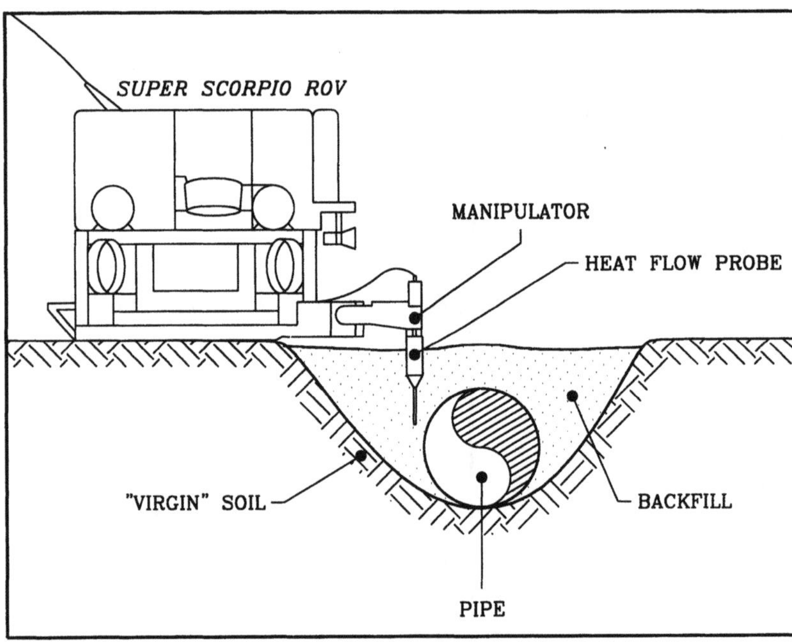

Fig. 9. In Situ Testing by ROV

3. ONSHORE TESTING AND ANALYSIS

On completion of the site work the sealed soil samples were transported to the Fugro soil testing laboratory in Hemel Hempstead U.K. and subjected to a rigorous programme of testing which included:

(a) Moisture Content

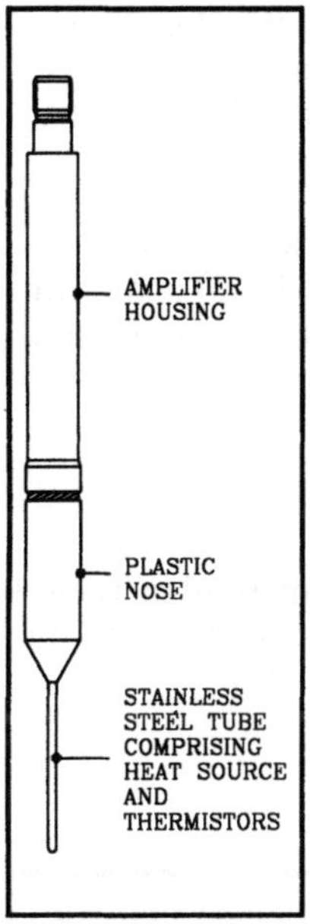

Fig. 10. Heat Flow Probe

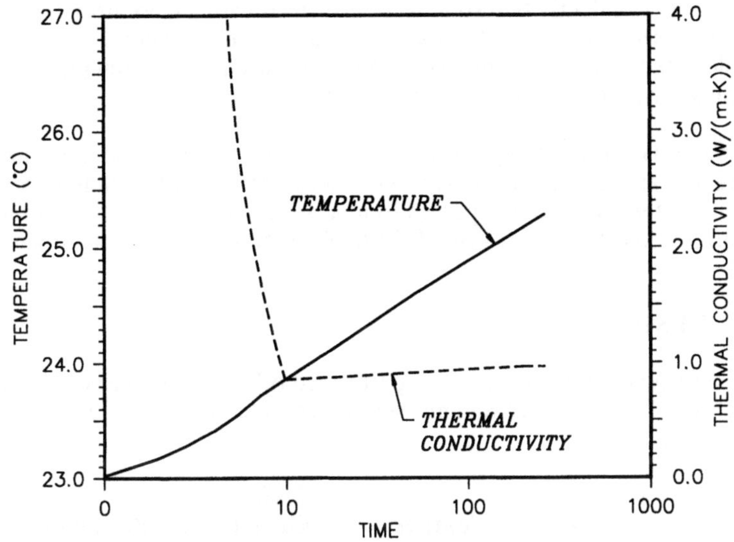

Fig. 11. Thermal Conductivity Test Result

(b) Atterberg (Liquid and Plastic) Limits

(c) Bulk and Dry Density

(d) Particle Size Distribution

(e) Specific Gravity

(f) Consolidation

(g) Thermal Conductivity

The purpose of the testing was threefold, namely;

(1) Confirm the similarity of the soil at the test locations to that along the Tordis-Gullfaks route.

(2) Determine the likely rate of consolidation of the back fill material.

(3) Establish with more confidence the likely range of Thermal Conductivity values for the soil.

The "Classification" tests - Atterberg Limits and Particle Size Distribution -, (British Standard 1377-1990) provided accurate "Finger printing" of the samples for comparison with the Tordis route soils.

Consolidation testing was performed by the "Constant Rate of Strain" method described by Janbu et al. (1981). Samples from near the top of the backfill were chosen to ensure that the effects of any self-weight consolidation that had already occurred would be negligible.

Thermal Conductivity testing was performed with a laboratory version of the Heat Flow Probe used during the site-work. In addition, thermal conductivity values were also calculated using the empiricial formula proposed by Kersten (1949) which relates thermal conductivity to soil type, moisture content and dry density.

Additional shear strength tests were not performed because the material was too soft to prepare as a Triaxial Compresison Test specimen (British Standard 1377-1990) and other simpler tests were not expected to give as reliable results as the laboratory and insitu tests performed offshore because of the likely effects of sample disturbance.

4. RESULTS

The results of most of the tests performed offshore and onshore are summarised in Table 1 below. The calculated and measured thermal conductivity results are also summarised on Figure 12.

TABLE 1
Comparison of Geotechnical Parameters

LABORATORY TESTS	TORDIS-GULLFAKS 'C' SOIL CONDITIONS		SNORRE-STATFJORD TRENCH BACKFILL MATERIAL
	MEDIUM DENSE to DENSE fine to medium SAND with shells (0.0 to 0.2-0.6 mBML)	VERY SOFT to SOFT CLAY with thin layers of sandy material (0.2-0.6 to 2.0 mBML)	VERY SOFT sandy CLAY with pockets and layers of fine to medium sand
MOISTURE CONTENT (%)	25-40	27-54	29-59
BULK DENSITY (Mg/m^3)	1.77-1.96	1.6-1.71	1.72-1.93
UNDRAINED SHEAR STRENGTH (kPa)	-	2-10	1-8
PLASTICITY INDEX (%)	-	10-25	5-22
THERMAL CONDUCTIVITY (W/(m.K))	-	0.92-0.96*	0.9-1.4

NOTE: (1) mBML - metres below mudline, (2) * denotes calculated parameter

As shown on Figure 12, the offshore thermal conductivity measurements all fall around 0.6 Watts per metre per degree Kelvin - W/(m.K) which is very close to the value for water (approx. 0.5 W/(m.K)). In hindsight it seems likely that the penetration was insufficient and the method of insertion of the probe, by means of the ROV manipulator, was not controlled enough to prevent remoulding of a zone around the probe. This may have significantly increased the localised moisture content and even possibly formed an annular space allowing water into direct contact with the probe. For future applications the insitu probe will be modified to facilitate deeper penetration and an insertion mechanism independent of the manipulator will be used.

In comparison the laboratory tests were performed under very controlled conditions and although the samples were underwater and fully saturated it was possible to ensure very good probe/soil contact by leaving the probe in the soil for longer periods before commencing the heating phase of the test.

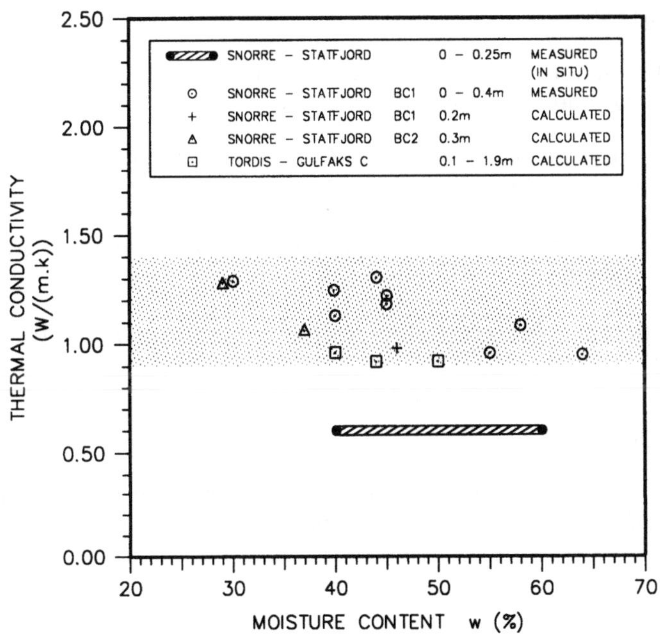

Fig. 12. Summary of Thermal Conductivity Results

The Consolidation tests and associated calculations indicated that the backfilled material should reconsolidate to close to its undisturbed condition in under two years.

The Influence of Other Factors on Insitu Thermal Conductivity

Consideration was also given to the possible influence of other factors that might affect the potential heat loss from the pipe namely: contact with virgin soil, temperature of the pipe, induced convection currents and natural seabed currents.

Settlement calculations indicated that the flowlines are likely to be in contact with "virgin" soil for over 40% of their circumference at the base of the trench. This will have the effect of reducing the thermal conductivity of the surrounding soil towards the lower bound of the predicted range (c.f. the calculated values for the Tordis route soils shown on Figure 12).

It is believed that the operating temperature of the flowlines will be around 40°C. The Heat Flow Probe equipment used in this investigation is not designed to operate above 30°C and therefore all tests were performed between 8°C and 28°C. Since no discernible temperature related trends were detected over this range it was considered reasonable to assume that this will apply between 30°C and 40°C.

Convection currents induced in the pore water of the backfill and surrounding "virgin" soil by the heat from the pipe are expected to be negligible because of their low respective permeabilities and at the water depths involved - 200 to 250 metres - seabed currents are not expected to have a discernible influence.

5. CONCLUSIONS

1. The soil conditions at the test sites were confirmed to be similar enough to those on the Tordis route to validate the exercise.

2. The shear strength and bulk density of the backfilled soil should be sufficient to counteract upheaval forces.

3. The thermal conductivity of the backfill would be low enough to give adequate thermal insulation.

4. The verification exercise therefore resulted in the acceptance of the "Revised Concept".

ACKNOWLEDGEMENTS

The authors are grateful to Saga Petroleum A/S, the other Tordis licence partners and Stena Offshore A/S for their permission to publish this paper.

REFERENCES

British Standard 1377 (1990) "Methods for Testing Soils for Civil Engineering Purposes".

Geise, J.M., Hoope, J. ten and May, R.E. (1988). "Design and Offshore Experience with an In Situ Vane," Vane Shear Testing in the Field and Laboratory, ASTM STP 1014, A.F. Richards, Ed., American Society for Testing and Materials, Philadelphia, 1988, pp 318-338.

Janbu, N., Tokheim, O. and Senneset, K. (1981), "Consolidation Tests with Continuous Loading," Proc. 10th Int. Cont. Soil Mech. and Foundation Eng., Stockholm, Vol. 1, Session 4, Paper 26.

Kersten, M.S. (1949). "Thermal Properties of Soils" Bulletin 28, Engineering Experiment Station, University of Minnesota, Minneapolis.

Zuidberg, H.M., Hoope, J. ten Geise, J.M. (1987). "Advances in In-situ Measurements", Proceedings of the Second International Symposium Field Measurements in Geomechanics, S. Sakurai Ed, Kobe, Japan.

NEW BURIAL TECHNIQUES FOR EFFECTIVE BURIAL FOR THE PROTECTION OF PIPES

I D BONNON
Technology Division
BT (Marine) Limited
Berth 203, Western Docks
Southampton, UK, SO1 OHH

ABSTRACT

This paper outlines recent developments which have been undertaken to improve techniques for effective cable burial and protection of subsea pipes.

Traditional methods of burial for subsea pipework have involved the use of towed plough systems, subsea dredging units, or mechanical or jetting based trenching systems. Recent vehicle developments as undertaken by BT (Marine) Limited (BT(M)) have embodied new techniques which will allow more efficient solutions for the pipeline owners. New vehicles now offer a hybrid of solutions depending on the local conditions which may prevail. For example, the new BT(M) Eureka vehicle has the ability to interface to four separate burial devices. There is a new advanced dynamic jetting system which maximises available power for efficient trenching. There are two separate mechanical tools; a rockwheel cutter for work in very hard seabed materials, and a chain excavator where very deep burial, in excess of 2m, is required. There is also a jet assisted plough which combines the benefits of water jetting with ploughed systems. All of these burial devices are controlled from the same host vehicle which is fitted with the latest sensing and tracking devices. Product handling is of prime importance and the new vehicle systems operate with either a non-product engagement system during burial, or an auto eject facility which ensures that in the event of hydraulic failure the product is safely ejected from the vehicle.

At present the burial vehicles have been designed to handle small diameter rigid pipes and flexibles. However, R&D studies have already proved the concept can be extrapolated for larger diameter pipes.

Volume 33: Aspect '94, 143–148.

INTRODUCTION

There are basically two methods by which pipes can be effectively buried in the seabed. One by the use of towed plough systems or secondly by the use of post lay burial devices.

The ploughs retain a very important role in pipe burial. The ploughs are particularly effective when long lengths of pipe have to be buried in difficult seabed conditions. Traditionally, the post lay burial vehicles would find it difficult to compete with the ploughs on a cost effective basis. However, recent advances in technological development have started to reverse this trend, and this aspect is discussed more fully within the content of this paper.

BACKGROUND

Many of the old post lay burial devices used for pipe burial suffered from three significant failings :

1. Problem with reliability
2. The potential for the mechanical cutting devices to damage the product
3. Could not be cost effective in comparison to the efficiency of a plough system.

Many of the older devices would also be very large and weigh, in some cases, in excess of 100 tonnes, making handling a key issue and requiring a large heavy work-class DSV to successfully launch and recover. By modern standards these early systems were very '*low tech*' and had relatively poor efficiency.

Recent advances have seen the introduction of new more efficient machines which have the concept of modular tooling operated from the same vehicle platform. The modular concept allows either a water jetting tool or a mechanical cutter to be fitted to the same vehicle and allow the vehicle to work more efficiently depending on the seabed conditions which are being encountered at the burial site.

OPERATIONAL TARGETS

The operational targets which must be met are as follows :

- The burial device must in no way damage the product it is burying.
- The post lay burial device must be able to find, detect and accurately track the pipe to be buried.
- The systems must be capable of working in a range of water depths, down to a maximum of 1,000m.
- The systems must have the facility to work on land or the beach if so required.
- The reliability of the systems must be such that it can be shown that a viable economic challenge can be made to some more traditional burial methods, ie plough systems.
- The systems should be able to bury exactly to the required depth of burial and have suitable sensors to confirm that burial has been achieved to the client's satisfaction during the burial process.
- The systems should be able to self verify the pipe has been buried to the designated depth in post survey mode.

BT(M) have already developed and built a vehicle system for the Eureka vehicle which satisfies all the above criteria. At this stage, the device is only targeted for the burial of cables, umbilicals and small diameter pipes. However, feasibility studies have already been undertaken which show that the same concept can be readily applied and expanded to cope with much larger diameter pipe burial. Obviously, there is always a penalty in terms of power budget, size and weight of vehicle, however, reasonable vehicle configurations can be established for tackling pipe burial in the most arduous seabed conditions.

Figure 1 below shows the outline of the Eureka post lay burial vehicle.

Figure 1 - Eureka

A MODULAR CONCEPT OF TOOLING FOR POST LAY BURIAL

The new generation of post lay burial devices should allow themselves to be interfaced to a variety of burial tool configurations. For example, on the BT(M) Eureka vehicle it is possible to interface to four separate burial tools.

Eureka has a Jetting Tool which utilises the dynamic concept of rotating jets. This Jetting Tool can straddle a product and effectively bury down to a depth of 1m. In addition, the Eureka vehicle also has fitted a Jet Assisted Plough. This device utilises a combination of self-generated drawbar pull on the vehicle in conjunction with a series of water lubrication jets for the plough share, this reduces plough tensions and allows a hybrid of jetting and ploughing to be effected.

To cope with the more difficult seabeds, Eureka is also fitted with a Rockwheel Cutter which allows burial in trenches down to 1.2m. This device is particularly useful for difficult sections where pipe burial is required through rocky areas or through boulder clay areas. There is a further tool which is also able to be interfaced to the Eureka vehicle, namely a Mechanical Chain Excavator which allows burial down to a depth of 2.2m. This Tool is particularly useful for beach landings or where there is a particular requirement for localised deeper burial.

By being able to interface all four burial tools from the same vehicle package, it is possible to go to a work site with one single vehicle solution. By interchanging tools, any seabed configuration or depth of burial criteria can generally be satisfied. Realising that product handling and safety is of paramount importance, the Eureka vehicle has also been fitted with one very novel concept.

If, at any time, during a vehicle operation when the Mechanical Rockwheel Cutter is being used, a total loss of power such as hydraulic failure or severed control umbilical, it is still possible to safely eject the product from the handling pathway through the vehicle itself. By activating an acoustic trigger a series of charged accumulators gradually and gently eject the product from the product handling pathway and it is safely deposited on the seabed, allowing the vehicle to be recovered without any damage being impaired on the product. This remote intervention feature has already been successfully proved on large diameter cable.

Although the Eureka vehicle is only presently configured mainly for handling telecommunication cables, large umbilicals and small diameter pipes and flexible pipes, concept studies have already established that the principle of the vehicle can easily be scaled to include the handling of much larger pipes in the future. Already, by applying the new advanced dynamic jetting concept, large diameter pipes can be efficiently buried to depth using jetting tools opposed to the traditional ploughed methods.

POINTERS FOR THE FUTURE

It is worth analysing the performance of some of the new cable burial machines as a pointer to what potentially could be achieved for pipe burial. Many of the old post lay burial cable ROV systems were specified with forward rates of progress of between 50-100m/hr in difficult seabeds. As new technology has become available, it has been possible to improve their rates of progression in some instances to 200-300m/hr forward progression, effecting burial in a single pass. During a recent operation in the North Morecambe field, undertaken by BT(M) using the Subtrak vehicle deployed off the Stolt Comex vessel, Seaway Harrier, it was possible to bury some 24km of cable in a single dive which only lasted 38 hours, to a specified depth of burial of 1m. At times the rate of progress was noted to be 1,500m/hr. Again, there is no reason why such performance cannot be scaled to cope with a much larger diameter pipe, given that the work vehicle itself will have to be accordingly scaled in terms of size, weight and power budget.

CONCLUSION

Recent achievements in ROV technology in the field of cable burial have resulted in more efficient and cost effective burial of subsea systems. Further advances in terms of making ROV systems more efficient, reliable and, more importantly, versatile to accept new advanced burial systems, should lead to further benefits. There now exists a huge potential to exploit many of the aspects of the cable burial technology field and project them for efficient burial of pipelines in the future. The concepts are proven, all that remains is to scale the processes of the new cable burial techniques for effective burial and then protection of pipes can become a reality.

HYDRODYNAMIC EXCAVATION - RECENT EXPERIENCE IN PIPELINE AND CABLE DE-BURIAL, TRENCHING AND BACKFILLING AND LARGE SCALE SEABED SITE CLEARANCE

N.V.SILLS
Underwater Excavation Limited
Badentoy Avenue
Portlethen
Aberdeen AB1 4YB
Scotland

ABSTRACT

"Hydrodynamic Excavation" is a phrase coined to describe the two part process of disrupting a soil's structure into its constituent particles and dispersing the resulting suspension in a controlled manner using a downward flow of water. The process is generally most applicable to non cohesive soils such as cobbles, gravels, sands and course silts and is specific to the underwater environment.
It's advantages in underwater operations such as pipeline and cable de-burial, trenching and backfilling have become apparent in recent operations, some of which are described in the paper.
The paper also summarises the development and operation of a series of machines," Jet Prop's", now used to exploit the process.

BACKGROUND

In the past, salvage companies have made use of the enormous underwater digging power of the flow of water from ship's propellers to refloat stranded ships and to uncover shipwrecks buried in the seabed. The latter was done by mooring the salvage vessel over a wreck site and placing a large 90 degree tube behind the ship's propeller to direct the flow at the seabed. However, this method usually referred to as "prop washing" is only useful in relatively shallow water due to the decay in water velocity of the produced water column with depth and required the vessel to be anchored.

To use the process with precision, in varying water depths, on different vessels and to monitor the excavation in real time requires considerable refinement of the technique and the addition of state of the art survey tools.

Volume 33: Aspect '94, 149–175.

The design and manufacture of machines capable of generating the necessary water columns and to operate effectively under varying parameters has required some particularly novel solutions.

CONCEPTUAL DESIGN

Research studies confirmed that the only viable method of generating the necessary low velocity, large volume, water column is by a propeller. As most people who have visited a ship's engine room will appreciate, the power required to drive even a small propeller involves large engines, gear boxes and shafts. The principle feature of a propeller is its relatively low RPM and high torque.

The first hydrodynamic excavation machines built, utilised high torque hydraulic motors to drive a propeller via an umbilical. Although successful in some applications they were too small to remove large volumes of soil quickly enough to make them cost effective in operations such as pipeline trenching. Much more powerful machines were required. Furthermore, preliminary design exercises showed that hydraulic or electric driven systems could not realistically be scaled up, where the power requirement was measured not in hundreds but thousands of horse power. Machines driven by these methods would be very large, heavy, require complex umbilicals and very expensive to manufacture.

It became necessary to invent an entirely new method of driving propellers. This concept utilised high pressure water jets mounted on the propeller blade tips as a drive mechanism. A novel, high pressure, water lubricated rotary coupling had also to be developed to supply the propeller. The resulting machine "Jet Prop" is now a patented device.

Although jet drives are inherently inefficient as a method of propulsion (and why most ships still have propellers driven directly by shafts), the advantages, when considered as a total system operating remotely in deep water, far exceed the disadvantages. No motors or gear boxes are required. The design is very simple, light and has very few moving parts. The power transmission fluid, water, is free, environmentally friendly and plentiful. Water, used as a power transmission fluid in an open circuit system is very efficient. Water has a low viscosity, compared to hydraulic oils and therefore less energy is lost

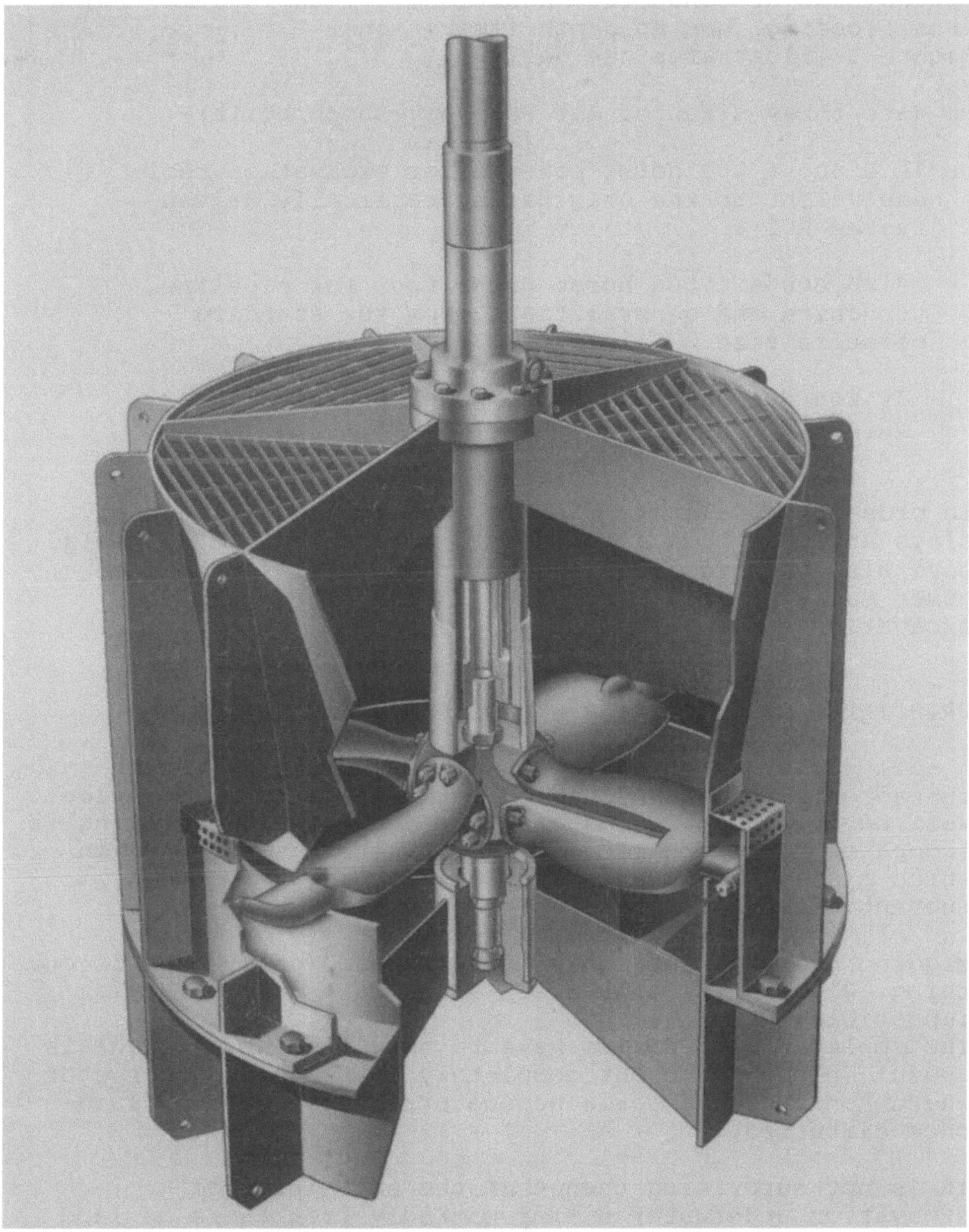

Figure 1: Internal Construction of Jet Prop 25,000 tool

in pipework. Also there is no necessity to have a fluid return hose or conductor. Additionally, the system, being free flooding, has no depth limitations.
Figure 1 illustrates the design.

To date three sizes of Jet Prop have been built:-

- JP 5,000 a 200 horse power diver excavation tool, equivalent to the original hydraulically driven system RUE).

- JP 25,000 a 1,500 horse power tool for pipeline trenching and general tasks. Now the standard offshore size.

- JP 250,000 a 15,000 horse power tool for mass excavation and presweeping.

In order to facilitate work in cohesive soils, such as clays and weak rock a series high pressure jetting tools have also been designed and built and utilise the same power packs and fluid conduits as the hydrodynamic excavation tools.

PRINCIPLES OF UNDERWATER HYDRODYNAMIC EXCAVATION

A soil submerged in water can exhibit quite different characteristics and mechanical properties to an identical soil on dry land. In addition the dense fluid overlying it (compared to air on land) entirely alters the manner in which a soil reacts when it is disturbed due to the buoyant effect of water.

Digging a steep sided hole in moist sand on a beach is one thing. Digging the same steep sided hole underwater is impossible.
The implements used also have to be different. A spade is useful on the beach but completely ineffective underwater where non cohesive soils become structureless and fluid once disturbed.

It is not surprising then that the principles of excavation underwater differ markedly from those on land and require a method specifically designed to operate in this environment.

The above illustrates an important aspect of the behaviour of non cohesive soils submerged in water. The soils structure is dependent almost entirely on gravity although over time other factors such as ice cover, wave action and precipitation of salts may have compacted, settled and bonded the soil.

Only where the particles are sufficiently small such as in the case of the colloidal sized particles of clays do the forces of electrical attraction take precedence and create a much tougher material.

Generally, if a non cohesive soil is disturbed sufficiently it will enter a fluid state. This is the first property exploited by the process of hydrodynamic excavation, the ability of a flow of water to stir up a soil into a fluid.

The second property exploited is that due to the specific gravity of water and its buoying effect. It has the ability to hold particles in suspension for short periods.

By imparting motion to the fluid the soil mass contained in it can be transported laterally across the sea floor easily. The process of hydrodynamic excavation utilises both these factors to advantage.

A large area of soil is disturbed into suspension at any one time and although the volume of soil may only form 3 to 5 percent of the total volume of fluid moved, this represents a very rapid excavation rate.

The figures denoting the size of Jet Prop above refer to the volume of water in cubic metres per hour generated by the unit at full power. Thus the 3 to 5 % of soil entrained within the volume of water generated by the Jet Prop 25,000 represents an excavation rate of 750 to 1,250 cubic metres per hour. The Jet Prop 250,000 being ten times more powerful can excavate at 7,500 to 12,500 cubic metres per hour. In practice precision is often more important than speed and lower than maximum powers have been used.

APPLICATIONS IN THE OIL AND GAS INDUSTRY

There are a large number of operations where hydrodynamic excavation has already proven to be a highly effective method, both within and without the oil and gas industry.

- Pipeline and cable trenching
- Pipeline and cable de-burial
- Pipeline and cable trench backfilling
- Pipeline freespan rectification
- Drill cuttings and cement slurry removal from wellheads and templates
- Rock dump removal
- Seabed preparation
- Local excavation for inspection

Other activities where the process should have a very significant impact but as yet has no track record are suggested below. These operations are of a scale that would normally require the utilisation of the 15,000 horse power Jet Prop 250,000 system.

- Sand wave presweeping
- Removal of iceberg scar ridges to assist pipeline installation. In some areas this would require the use of the system's ancillary clay cutting tools.

SYSTEM COMPONENTS

The systems are all designed to be self contained and operable from a variety of surface vessels.

Each system consists of:-

- Underwater excavation unit
- Deployment system
- High pressure water pump(s) and associated pipe work
- Control container
- Sonar and TV system

The type, scale and depth of the intended operation dictates the configuration, size and ancillary support mobilised. For most oil and gas field applications standard size JP 25,000's are used. This tool is extremely robust, weighs only 2.5 tonnes, yet can operate at up to 1,500 horse power.

In certain operations where a single standard JP 25,000 has insufficient capacity, two or more machines can be used, either bolted together or individually in-line one behind the other. This latter configuration can be used where deep trenches are required but pipe stress limitations require a long transition from seabed to trench bottom.

For operations deeper than 300 metres or where the work is undertaken from a drillrig the JP 25,000 is generally deployed on drillstring.

To allow a high level of compatibility with existing offshore technology and practice and ensure a large pool of equipment is readily available, standard drilling industry pumps, (triplex slush pump type) hard iron and flexible pipe work, control valves and fittings are used to power the excavators. Use of drilling industry equipment and practice also allows the employment of highly trained and experienced drilling personnel on offshore operations with all its attendant benefits to health and safety at work.

OPERATION OF THE SYSTEM

The system is mobilised to a vessel, commissioned and once the vessel is over the work site the tool is deployed to the appropriate height above the ground. The on board sonar and TV systems are used to watch down and position the unit.

The principle function of the underwater excavation unit is to generate a large diameter, low velocity water column and direct it vertically downward at the work site. It also acts as a platform to carry the various monitoring and survey systems necessary for controlling and recording the excavation activity.

Control of the excavation activity is achieved by adjustment of three parameters. The height of the unit above the seabed or work site, generally between 3 and 8 metres in the case of the JP 25,000 and between 8 and 20 metres for the JP 250,000, power setting and speed over the ground. Speed over the ground is controlled by moving the surface vessel on DP or with anchors. It is therefore important for the vessel to have an accurate position fixing system.

The column of water projected at the seabed is used to both fail the surface of the soil and to carry the resulting suspension horizontally in a radial pattern to a distance according to requirements. At high powers the same system can be used to fluidise suitable soils to a predetermined depth and let pipelines and cables sink.

Due to the fact that the water column generating tools are relatively small (JP 25,000 is 1.7m high and 1.8m in diameter) and light and operated many metres above the work site, there is very little risk of contact damage to sensitive coatings or structures.

The low velocity of the water column, generally between 3.0 and 6.0m/sec for trenching, is insufficient to cause damage to even the most fragile underwater structures. In addition the system can be deployed, operated and recovered safely in weather conditions far above that of conventional seabed orientated systems such as tractors and ploughs.

A typical pipeline trenching operation would require the unit to be stationed 3 to 4 metres above the pipeline and maintained within a horizontal accuracy of plus/minus 1.5 metres. Heave due to the motion of the vessel is generally acceptable up to 4 metres and has only a marginal effect on trench profile.

With the vessel stationary the power is gradually increased until the excavation is observed to be initiated (on the sonar) and then the appropriate vessel move begun. Sonar and TV monitors are installed at the vessel control station to allow the vessel operators to follow the excavation or trenching process.

The trenching personnel set and control the profile and depth of the trench from the control cabin and request the vessel to proceed at the appropriate speed, in most operations this will be between 60 and 120 metres per hour for a trench depths up to 2 metres. For deeper trench depths speeds will be lower, for shallower depths speeds will be higher. Where commercial conditions demand more rapid progress multiple excavators can be used.

Hydrodynamic excavation confers a considerable flexibility on operations as the trenching or other excavation activity can be suspended and restarted at will and progressed close up to subsea structures. The process is versatile and can be used to de-bury a trenched pipeline or cable, backfill trenches in many circumstances, remove rockdump or deepen an existing installation for such purposes as freespan correction.

CONTROL AND MONITORING

An important aspect of any trenching or general large scale excavation is the ability to follow progress in real time and record results with great accuracy.

In the case of the Jet Prop system this is achieved by installing a high specification real time sonar profiling system on the excavator itself. Reson Systems Seabat 9001 and 9002 systems are routinely used. This device is capable of taking 30 profiles a second and can effectively "freeze" a profile even when the excavator is heaving up and down. Unlike other types of sonar this rapid up date of profiles allows it to "see through" the suspended material. Although it uses 60 discrete 1.5 degree conical beams arranged in a swathe, it is has also proved capable of identifying even small pipelines and cables readily as the distance from the target is relatively small.

This allows the control personnel (and also a DP or anchor winch operator) to maintain contact with and therefore follow a pipeline throughout the trenching operation.

Although at present these facilities are used only to control and monitor ongoing activities, the final surveys being conducted later by ROV's, it is planned to incorporate a very accurate positioning system based on Seatec's innovative "Smart Wire" system into the spread. This, when used in conjunction with a global positioning system installed on the surface vessel, will provide absolute position and attitude of the Seabat sonar in relation to a known datum such as LAT. This will give the ability to conduct fully quality controlled surveys from the excavator before, during and after trenching. It is expected that this system will cut costs significantly as it will remove the requirement for a survey ROV spread.

This system is being developed as recent experience has shown that surveys based on existing seabed and through water acoustic positioning systems, as used in ROV surveys, are prone to down grading and intermittent repeatability when high levels of suspended soil and acoustic noise are present. However, the author is aware of at least one development in acoustic position fixing that may improve this situation.

During operations the Jet Prop unit draws water through an intake many metres above seabed level and so effectively floods the excavation site with clean water. A SIT camera installed in parallel with the Seabat sonar (and looking directly down) is therefore most often able to visualise a pipeline or cable within its view. Continuous visual inspection is thereby available .

The Aspects '94 conference presentation will illustrate these aspects with video and sonar recordings.

CASE HISTORIES OF OPERATIONS

The following case histories give examples of recent operations in a variety of fields and applications. Operations using the small diver operated JP 5,000 system are omitted.

Case 1. De-burial of a 400m section of live 32" gas pipeline

- Bruce Field
- Client, Comex
- Water depth, 130 metres
- Soil conditions, natural sandy silty backfill.

A single JP 25,000 deployed from Comex's semi-submersible DSV "Uncle John" was used to completely expose a 400 metre long section of Total's FUKA pipeline. This was required to allow the pipeline to be lifted during connection of the Bruce pipeline by "hot tapping".

The FUKA pipeline was originally trenched to a depth of 3 metres by Jet Barge some 11 years prior to the de-burial operation. Natural backfill consisting of a soft sandy silt had completely infilled the 5 to 6 metre wide trench to a degree that the trench was completely obscured. The surrounding natural seabed comprised of dense medium to fine sands interspersed with shell layers and gravels.

The objective was to preferentially remove all backfill material from the trench without disturbing or expanding the original trench. The total soil volume to be removed was calculated to be between 5,000 and 6,000 cubic metres.

In order to locate the pipeline the tool was deployed to 4 metres above the seabed. A short length of trench was cut, at half power, across the estimated track of the pipeline by manoeuvring the vessel on DP. A diver was used to locate the pipeline and guide the vessel over the exact position of the pipeline. This procedure was used as at that time no Seabat sonar was fitted.

The machine was then powered up to 35 % power (450 horse power) and the vessel manoeuvred along the track of the pipeline and checked by diver periodically for position. The excavation operation took 18 hours to complete with an average excavation rate of 300 cubic metres per hour. An ROV survey conducted after the operation showed that in all except a very few places the pipeline was completely exposed and that the walls of the original trench were intact.

See figure 2. Sonargraphic survey of part of trench. The "3D" sonargraph shows approx 65 metres of trench. The pipeline is not visible due to the sonar data processing technique where "anomalies" are removed!

Case 2. Excavation of natural seabed to install "Hot Tap" frame

- Bruce Field
- Client, Comex
- Water depth 130 metres
- Soil conditions, medium to fine dense sand with interspersed shell and gravel layers.

This operation was a continuation of the above Case 1. work and from the same vessel. The pipeline remained under pressure and in service throughout the operation. Here a single JP 25,000 was deployed to excavate an area 35 metres long, 15 metres wide and 4 metres deep but in

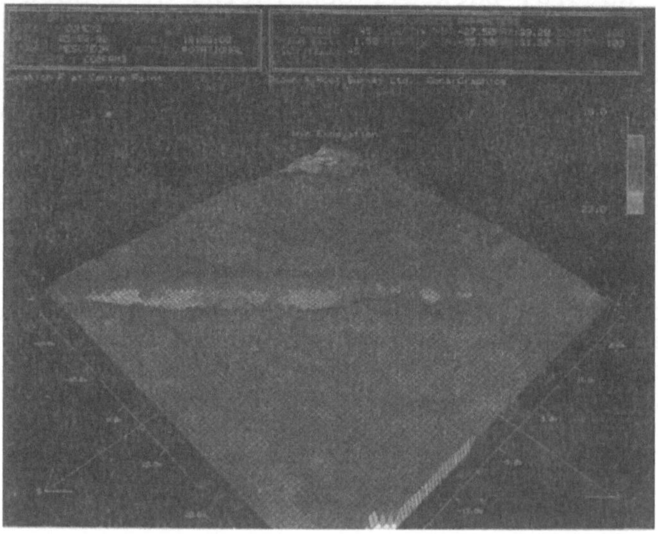

Figure 2: Sonargraph of 65 metres of trench (courtesy British Gas Ltd)

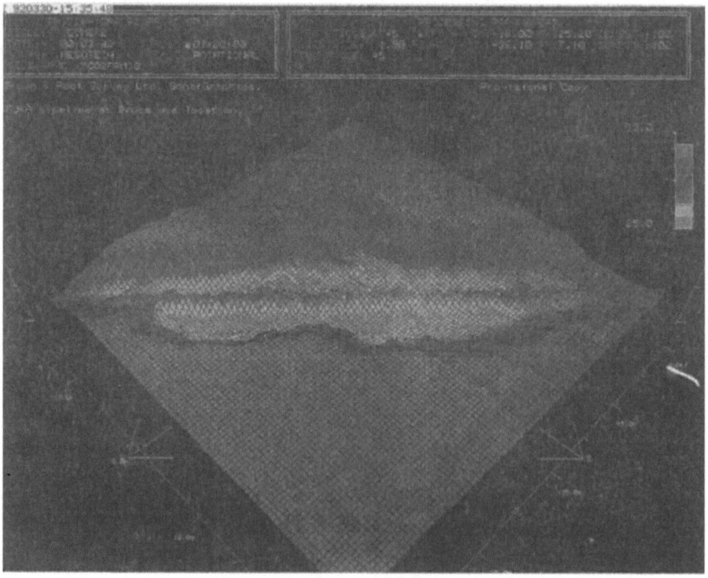

Figure 3: Sonargraph of excavation for "hot tap" unit. Pipe line appears 2 metres high due to processing (courtesy British Gas Ltd)

natural seabed rather than the disturbed soil of the trench. This was to allow the installation of a large " Hot Tap" frame.

The machine was manoeuvred 25 metres along both sides of the pipe and 5 metres to each side. The previous pipeline deburial operation had removed a central corridor 6 metre wide and 3 metre deep. Other parameters used were 50 % power and a 3 metre altitude above seabed. The operation took 6 hours in which time some 1,500 cubic metres of seabed was removed.

See figure 3. Sonargraphic survey of the area post excavation.

Case 3. Freespan correction on 24" Sliepner pipeline

- Sliepner Field
- Client, EMC Ltd.
- Water depth 276 metres.
- Soil conditions, soft clay.

A JP 25,000 system with Seabat sonar and SIT camera/lights was mobilised to the EMC's DSV" Bar Protector" to correct a 120 metre long freespan. The 24", air filled, pipeline crossed a 4 metre deep, 120 metre diameter "pock mark" in 276 metres of water.

Previous efforts to install a 4 metre high rock "sleeper" to support the pipeline had been unsuccessful due to the low soil strength in the centre of the "pock mark" and the dispersal of the rock over a wide area due to depth. The 8,000 tons of rock dumped had either sunk into the soil or landed on the shoulders of the "pock mark".

To allow the pipeline to be flooded for testing it was necessary to reduce the freespanned length from 120 metres to 60 metres or less. This required lowering the centre of the freespanned pipeline by 2.5 metres. To avoid over stressing the pipe required two 300 metre long transitions, one either side of the "pock mark". In addition, to prevent any further increase in the already critical span length the entire 720 metre section had to be lowered in 25 centimetre steps.

The vessel's crane was used to deploy the Jet Prop and a

Figure 4: Jet Prop 25,000 deployed from "Bar Protector" crane to 276 metres

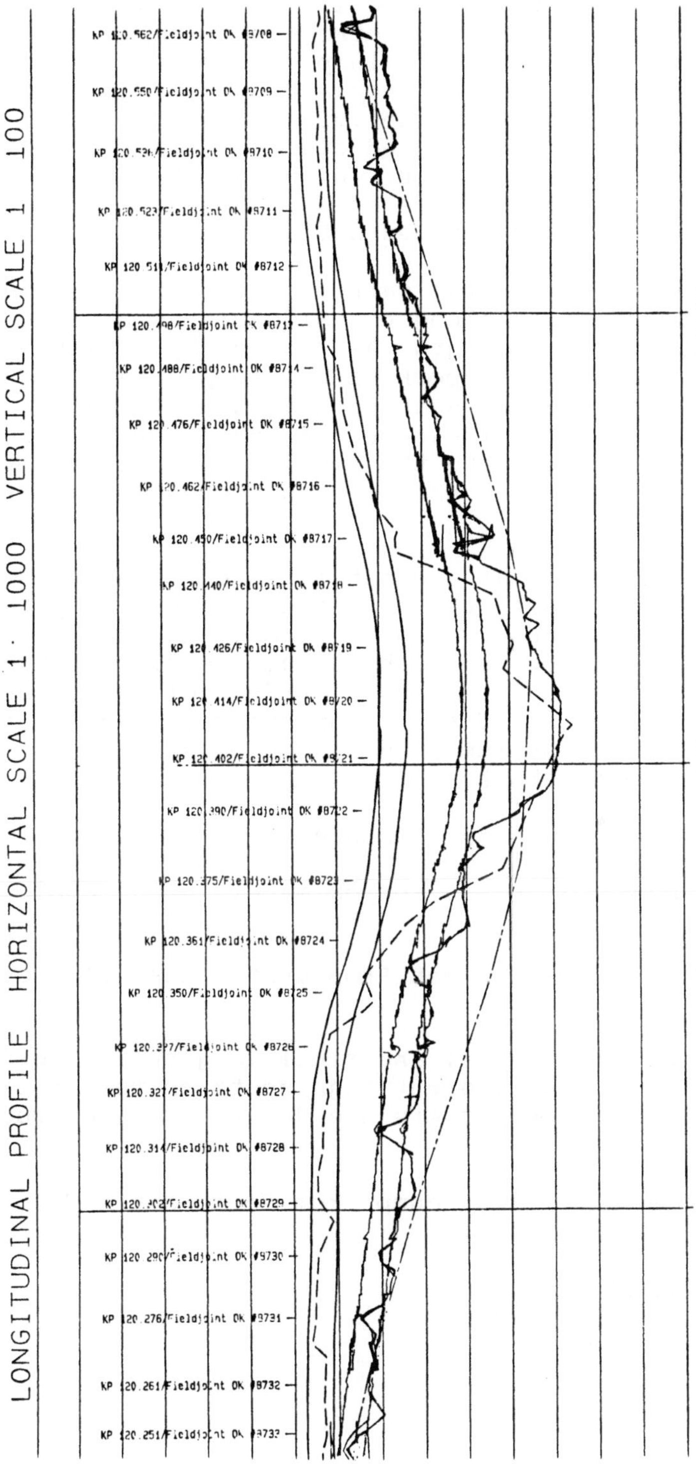

Figure 5: 300 metre section of longitudinal profile through pock mark, prior to flooding and testing pipeline

separate winch used to deploy a 1 ton clump weight attached to the unit by a 10 metre long strop. The two were deployed in tandem. The clump weight was used to ensure alignment of the Jet Prop and thereby the profiling sonar and TV systems, with the vessel heading. Once deployed the unit was navigated over the start position using the unit's Seabat sonar, which was able to identify the pipeline from an altitude of 200 metres. Pre and post trenching surveys were carried out using an ROV.

The trenching operation was carried out by making ten 25 centimetre deep passes along the side of the pipeline. Each consecutive pass length was reduced by 60 metres (30 metres each side of the "pock mark") to provide the required transition. The Seabat sonar and SIT camera were used to control depth of excavation and record progress. In order to navigate along the pipeline, remote sonar and TV camera monitors were installed on the bridge in front of the DP operators.

The vessel was driven at a constant speed along the track of the pipeline keeping the pipeline in the centre the sonar profile (by manoeuvring the vessel laterally on DP as required). This proved to be a surprisingly accurate method due in part to the damping effect of water depth but also to the skill of the DP operators. In all but one instance the accuracy remained within plus/minus 0.5 metres of intended track.

Total excavation time was 60 hours.

Total excavated volume was approximately 3,000 cubic metres.

Figure 4 illustrates the Jet Prop system installed on "Bar Protector".

Figure 5 shows the initial and final longitudinal profiles of the pipeline. The apparent high spots on the profile are largely anomalies, others are due to areas of rock dump.

Case 4. Removal of rock dump from Cormorant - UMC 8" pipeline

- Cormorant Field
- Client, Shell Expro

- Water depth 160 metres.
- Material, 8 to 12 inch granite rockdump.

This operation was required to remove a 20 metre section of rockdump overlaying several 3 metre high, 7 metre long concrete pipeline covers. Depth of rock dump over the covers varied from 1 to 3 metres and extended approximately 3 metres to either side.

The operation was conducted from Shell's MSV "Stadive" to allow the covers to be removed and access for divers to disconnect the damaged pipeline.

A JP 25,000 system was deployed to a position 3 metres above the rockdump and run along the centre line of the covers at 1,250 horse power. Two runs, one either side of the covers were then made to throw the rock to the sides of the covers. Including deployment and recovery the operation took 8 hours.

Case 5. Cable repair, Location, de-burial, trenching and backfilling

- Rolf field
- Client, Stena Offshore Ltd
- Water depth, 35 metres
- Soil conditions, very fine dense sand.

Following the failure of this 6,000 volt 4" platform power supply cable and an unsuccessful 48 hour ROV and diver operation to locate the buried cable a JP 25,000 was deployed from the DSV "Stena Orelia" to locate and debury the failed section.

The tool was deployed to 4 metres above the seabed and directly over the estimated position of the cable. This was done, rather than cutting a trench across the track of the cable to intersect it, due to the proximity of a small diameter pipeline. A two metre deep, 5 metre diameter spot excavation was made at low power. This excavation took less than one minute with the cable being visible on SIT camera and sonar. Cable depth was measured to be 0.75 metres below seabed.

The cable was then de-buried for 50 metres either side of the initial excavation, cut and one end retrieved to the surface. This operation took 6 hours and was slow due to poor weather conditions and some difficulty in following

Figure 6: Single Jet Prop 25,000 with Seabat Sonar and SIT camera fitted. Stena "Orelia"

the small cable on sonar. Although excellent during day light hours the SIT camera observational capability during night time was poor due to back scatter from the lighting system. Particular care had to be exercised to prevent de-burying the adjacent pipeline.

Due to water ingress into the cable, further lengths had to be de-buried and cut out. In total 0.5 kilometres of cable was deburied. To ensure the minimum stress was placed on the cable during recovery of the ends to the surface vessel, it was important that the cable was completely uncovered. This was achieved by trenching at very low power and speed (200 horse power at 30 metres/hour) and maintaining the trench depth at 1 metre and 5 metre width. After several hours the trench would stabilise at 0.75 metre depth and 6 metres width due to slumping of the trench walls. (Cable was buried 0.75 metres in very fine "sugar" sand).

This additional 400 metres of de-burial took a total of 13 hours.

The cable was repaired by first raising one end on deck and splicing in a 500 metre length of new cable. This was paid out onto the seabed as the vessel manoeuvred back to the other end. This second end (which had been previously capped and laid back on the seabed) was recovered to deck and spliced to the new cable. The cable was laid down on the seabed.

The Jet Prop was then positioned at one end of the repair at 2.5 metres altitude and run at 65 % power to ensure a steep sided trench. A 2 metre deep trench was cut along the exposed 0.5 kilometres of cable. Due to the presence of several tight curves in the cable, the operation was undertaken at 60 metres/hour speed to allow the vessel adequate time to manoeuvre. The operation took 10 hours to complete. The final as built ROV survey showed that the 2 metre deep trench had collapsed as expected and backfilled to give an average cover over the cable of 1 metre.

During high voltage testing the cable failed 4 kilometres further up the cable and the whole operation was repeated in a force 8 gale.

Figure 6 shows the Jet Prop ready for deployment.

Case 6. Trenching trial on a redundant 12 inch pipeline

- Dan and Skjuld fields
- Client, Stena Offshore Ltd
- Water depth, 40 metres
- Soil conditions, very fine dense sand

Stena Offshore Ltd conducted a series of trials from their DSV "Stena Orelia" in order to establish and document the performance of both single and dual Jet Prop 25,000 systems. The aim was to cut a series of trenches to different depths using different parameters of power altitude and speed over the ground. To survey the results a high specification ROV survey spread was also mobilised including the facilicty to produce bathymetric charts offshore.

The redundant pipeline between the Dan B Platform and the decommissioned PLEM site was known to be buried.

On arrival at site a JP 25,000 was deployed and two cross trenches cut, one 20 metres from the abandoned PLEM and one 200 metres from the Dan B platform. The ROV took accurate position fixes of the exposed pipeline. These were used to compute the track of the pipeline and an ROV survey of the entire pipeline route made as a control. This showed that the 1.3 kilometres of pipeline between the two points was completely buried to a depth of 0.75 metres TOP as later discovered.

In order to allow the DP operators to navigate the vessel along the pipeline, sonar and TV monitors were positioned in front of the DP console. This allowed the operators to maintain lateral position over the pipeline. It had been planned to use the satellite based Geographical Positioning System backed up by shore based reference "Differential" stations (DGPS) for providing constantly updated longitudinal position data to track follow the pipeline. In the event this proved to be unstable and could not be used for this purpose.

It was therefore decided to use the vessel's two taut wires to "walk" the pipeline by dialling in offsets. In the shallow water on site this meant picking up and redeploying the taut wire every 10 metres and lead to a "jerk" each time a transfer from one taut wire system to the next was made.

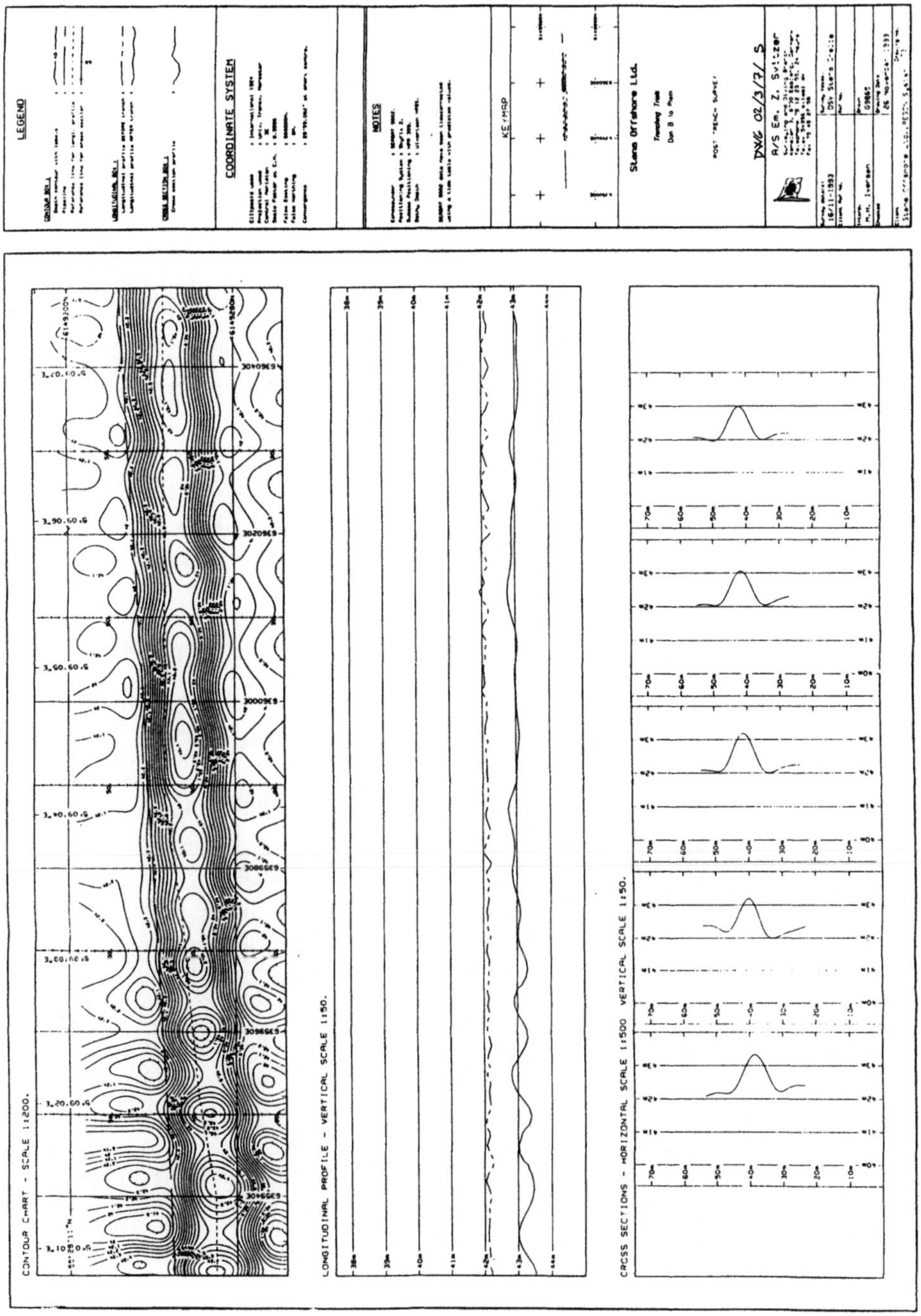

Figure 7: Effect of taut wire charger over on trench shape. Pipeline is 0.3 to 0.5m below bottom of trench backfill

Figure 8: Dual Jet Prop 25,000 with Seabat Sonar and SIT camera fitted. Stena "Orelia"

Figure 7 illustrates this effect on a 120 metre section of trench. Top of pipe is 0.3 to 0.5 metres below trench backfill in the centre of the trench.

Prior to starting the trenching trial the pipeline had to be dug up. A 1 metre deep de-burial pass was therefore made at 120 metres an hour, 5 metres altitude and 700 horse power. The objective of the de-burial was to excavate as wide a trench as possible to simulate a post pipelay situation on flat seabed. The subsequent ROV survey showed the trench to vary between 10 and 12 metres wide. In retrospect the assumption that this represented a postlay situation on natural seabed was misleading since the soil had been reworked.

An important aspect of the trenching trial was to leave the pipeline exposed after each pass to allow the ROV to make an Out Of Straightness survey (OOS). For this purpose the UDI Scorpio ROV was fitted with a set of wheels, which allowed the vehicle to maintain a constant known distance off the pipe and an odometer to record KP's.

The first trial trench was made to a depth of 1.75 metres at a speed of 60 metres/hour which was achieved after interruption by a period of bad weather and diversion to an emergency cable repair.

A second trial using a single Jet Prop to excavate a 2.5 metre deep open trench was successful but immediate collapse of the trench walls after the passage of the machine caused considerable backfill and prevented OOS surveys using the ROV odometer system.

A third trial was therefore conducted using two machines bolted together. This gave the desired trench depth with ease at speeds of 60 metres/hour but due to the fact the soil had been reworked several times it proved impossible to prevent the trench backfilling within a few hours and burying the pipe. Whilst this was seen as undesirable in the trial because it prevented gathering OOS data in many other operations, trenching and backfilling in one pass would be viewed as a benefit.

Figure 8 shows dual Jet Prop prior to deployment.

A final trial was conducted to demonstrate the ability of the system to actively backfill a trench by manoeuvring the tool along one side of the trench and blowing the spoil heap and local soil back into the trench. This proved a simple and effective procedure despite being undertaken in a force 9 gale. On average 0.7 metres of soil was deposited over the trial length of 360 metres at 120 metres/hour. Depth of cover data was gained using a TSS 340 pipetracker.

Postscript. Since the above trials a new procedure for trenching in this type of soil and to allow accurate determination of TOP to seabed level and depth of cover has been adopted.

See figure 9.

In this procedure a second Jet Prop is deployed just ahead of the pipeline touch down point in the trench, to locally re-expose a short section of pipe. This allows measurement of the TOP to seabed distance using a Seabat sonar. The measurement is made either by screen dumping selected profiles through a lap top computer and scale measurement from the hard copy print or by "freezing" a profile on the sonar monitor and using the screen cursor to take direct measurements.

Case 7. Pipeline Tunnel Piercing Point site preparation.

- Kollsnes, Norway, Troll Project
- Client, Rockwater A/S
- Water depth, 170 metres
- Soil conditions, Moraine, soft clay, boulders and cobbles.

This operation required the 15,000 horse power Jet Prop 250,000.

The system comprising of 600 tons of equipment was mobilised to the Rockwater "SEMI 2" to clear three 40 metre diameter areas of loose material including several hundred large boulders. In addition the central 20 metre diameter area of each site required the removal of 1 to 2 metres depth of morainic clays and cobbles and cleaning of the underlying sheet rock to a finish that would allow detailed examination of its structure. ROV surveys were required to locate and measure any fissures or cracks in the rock structure to access the rock's integrity. Pipelines, associated with the Troll field, pass vertically downward through drilled and blasted vertical shafts and into tunnels for passage ashore. This work was required to enable the final breakthrough to seabed to be made with confidence.

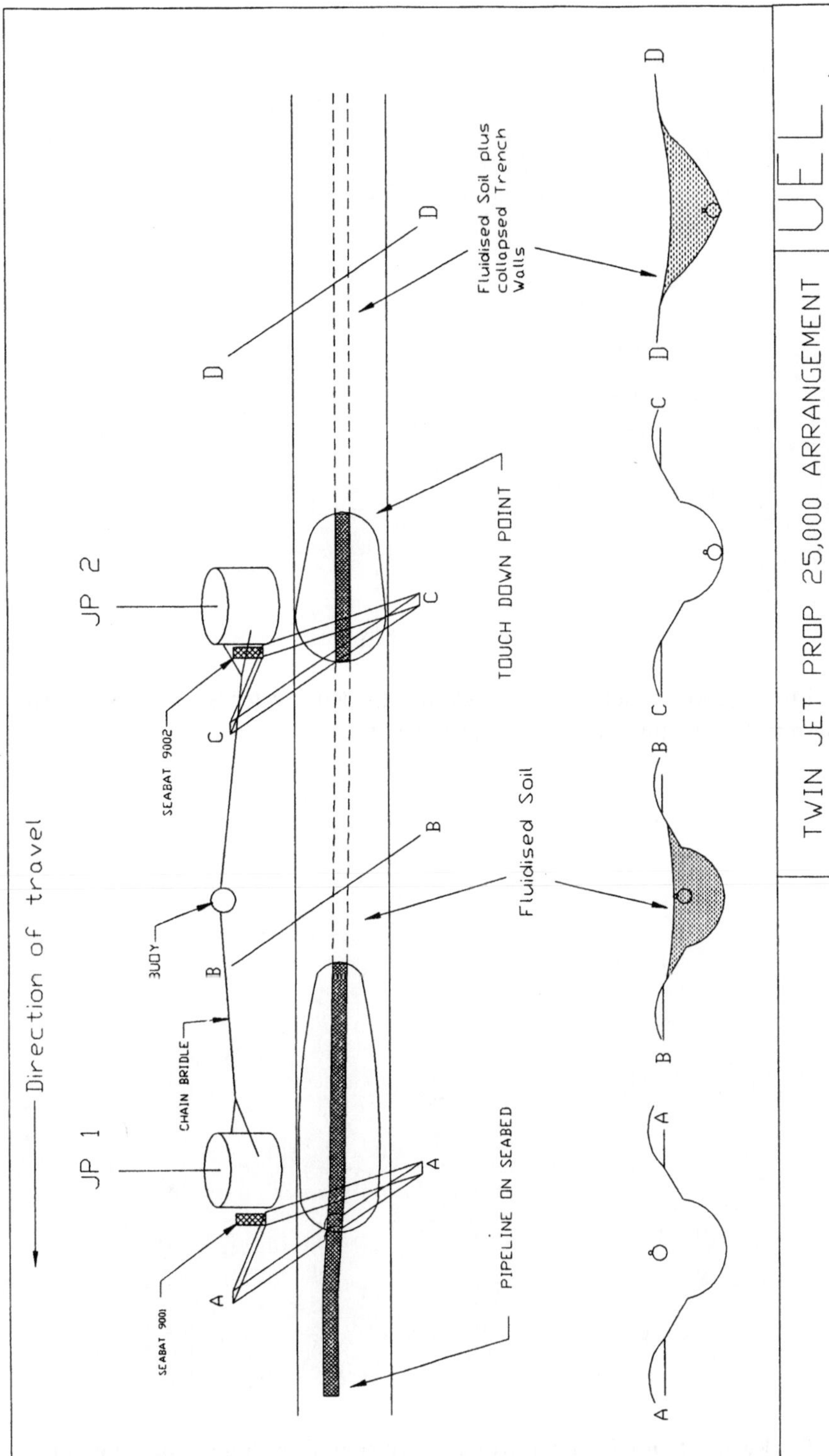

Figure 9: Sketch of trenching procedure using 2 Jet Prop units spaced 25 metres apart

Figure 10: Mobilisation of Jet Prop 250,000 aboard Rockwater Semi 2. Note drill rig handling system to left, pumping to spread to right

Figure 11: Rockwater Semi 2 on site Kollsnes, during Troll tunnel piercing site clearance excavation. Note drill rig

The SEMI 2 was stationed on DP centrally over the first site. The Jet Prop was deployed to 12 metres above seabed on 6 & 5/8" drillstring and powered up to 10,000 horse power. The vessel was then manoeuvred out to a 10 metre radius and a complete circuit made around the central point. A second circuit was then made at a 5 metre radius to ensure the rock surface was thoroughly cleaned. The overall procedure took 8 hours and was repeated at each of the three sites. ROV surveys showed all loose materials to have been expelled to a radius of 20 metres. Small areas of tenacious morainic material remained in isolated places and were subsequently removed in a further 12 hours of remedial work.

Figure 10 & 11 illustrates the Jet Prop 250,000 system on the SEMI 2.

Session 4
Operations and Maintenance

THE INERTIAL GEOMETRY PIG

J.BRUCE
Nowsco Pipeline Services
Howe Moss Drive
Kirkhill Industrial Est.
Dyce, Aberdeen. AB2 0GL
Scotland

ABSTRACT

Operational experience with pipelines around the world has shown that the main causes of pipe failure are related to changes in the geometry of the line. These changes typically include subsidence, freespanning , crushing of the pipe, and other instances of external damage, and have been assessed wherever possible, in conjunction with the original as laid condition of the line to provide for change monitoring. Structural deformation monitoring and measurements of the in situ geometry of new or existing pipelines provides input to pipeline reliability systems that ensure the most economic operation of the pipeline.

The **GEOPIG** is an intelligent survey tool which is run in the pipeline as a conventional pig, and which collects data on the pipeline during its traverse. The results of this survey can be used as a basis for fitness for purpose determinations, or in critical assessments, can be enhanced by overlaying it with corrosion (wall thickness) data to provide a complete picture of pipewall, profile and condition. Apart from its obvious application to pipeline inspections, the tool has been used for more diverse applications related to commissioning, laying of rock dumps, mapping pipeline routes, and assessment of subsidence, scouring and upheaval, internal calipering and baseline surveying of pipelines. To facilitate interpretation, a software analysis system has been developed.

The paper describes the two database systems which have been developed to handle and display **GEOPIG** data, and presents a range of **GEOPIG** data to illustrate the various detection capabilities of the tool. **GEODISPLAY** is a menu driven software package which utilises an interactive database system. It displays both the raw data collected from the pig as well as the comprehensivly processed data. **GEODENT** is also a menu driven software package, but is used to display and analyse the processed data relating to suspected dent and their location on the pipeline.

Volume 33: Aspect '94, 179–193.

INTRODUCTION

The basic integrity of a pipeline may be assessed by measuring a small, but critical number of parameters relating to it, at a suitably close measuring pitch, these include,

1. Pipeline centre position and attitude in space.

2. Pipeline cross section.

3. Remaining wall thickness.

The **GEOPIG** tool has been developed to provide a significant amount of this data in a single pass through the pipeline system. The **GEOPIG** is designed to determine position, orientation, curvature, bend radius, displacement, subsidence, bending strain, ovality, and deformations (such as dents and wrinkles). The heart of the device is a strapdown inertial measurement unit (SIMU), which produces a three dimensional measurement of inertial acceleration and angular rate directly from orthogonal triads of accelerometers and gyroscopes. The SIMU is coupled via a processing unit to provide a continuous sensing of position and attitude. The positional data is combined with ultrasonic caliper measurements of the pipe profile taken by annular arrays of ultrasonic transducers mounted on each end of the inertial system. Combination of these sonar readings translates to the pig's pitch, and heading to the slope and azimuth of the pipeline. The data on the distance travelled is obtained from odometer's in contact with the pipewall cross referenced by weld detect sensors also built into the tool. The **GEOPIG** is completed by a tracking transmitter for location, and a storage device and power supply which allow independent operation for long measurement periods.

On completion of a run through a pipeline, the tool contains a comprehensive set of data that describes the path and cross sectional history of the line, which can be processed by reporting software into a variety of formats useful to engineers assessing the line condition.

Figure one and two are schematics of a 48 inch **GEOPIG** and a 12 inch **GEOPIG** respectively.

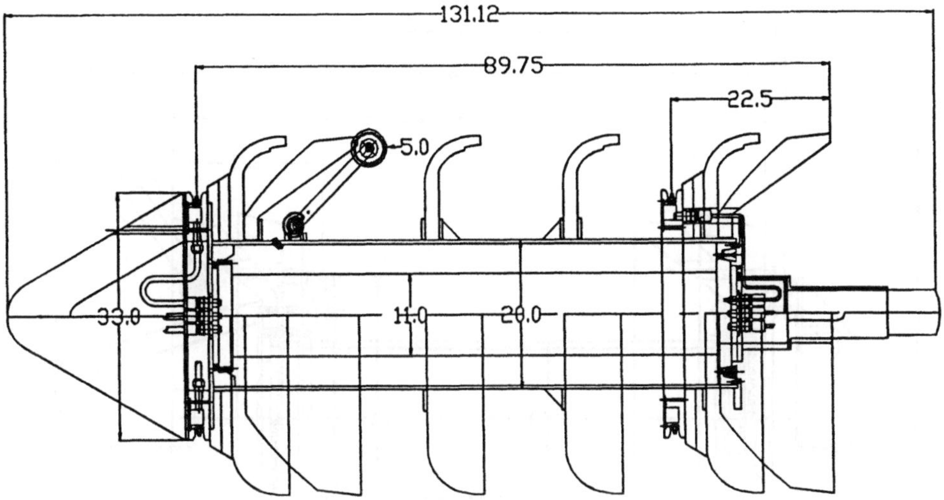

Figure 1
48 Inch Oil Geopig

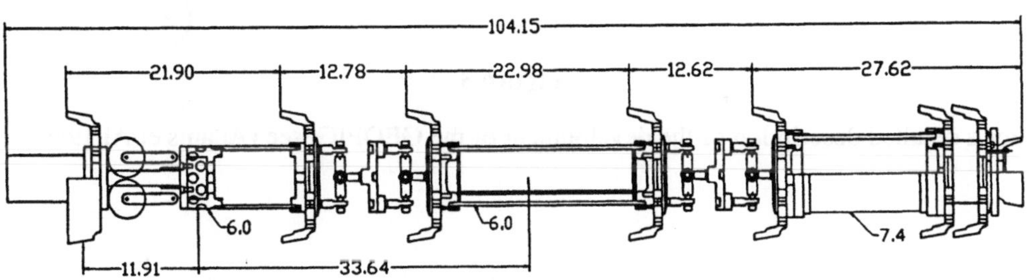

Figure 2
12 Inch Oil Geopig

Figure three is a schematic of the **GEOPIG** for NPS 28 inch size and larger.

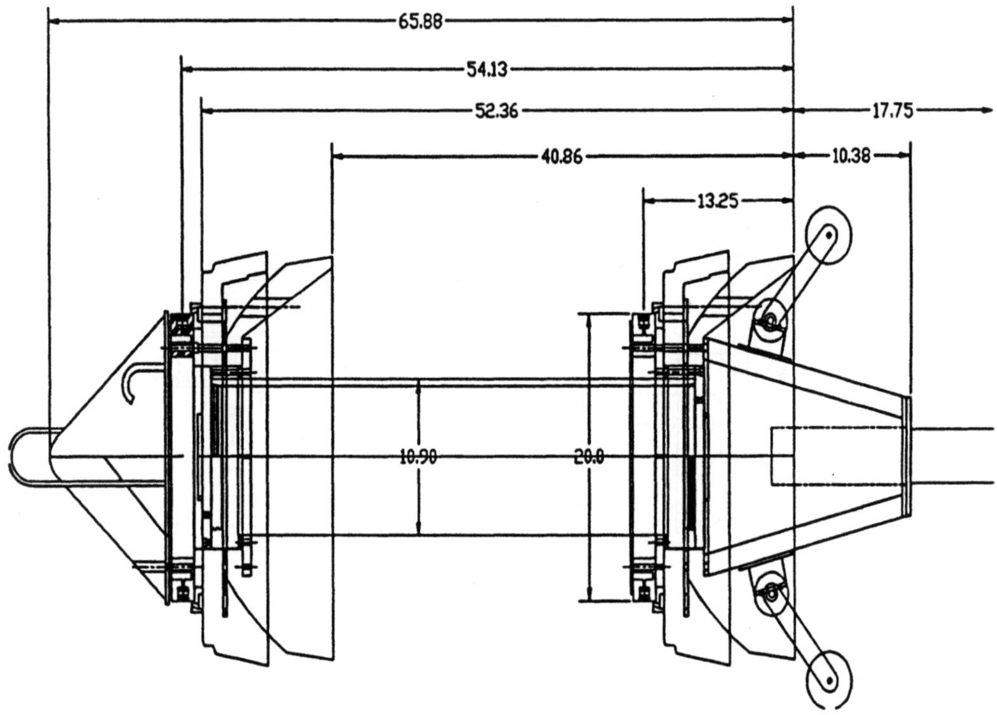

Figure 3

For a detailed description of the development of the GEOPIG, see (Adams et Al 1989).

BACKGROUND FOR COMPARISON SOFTWARE

In the initial stages of development delivery of data was in hard copy form. A typical report would consist of several 3 ring binders. For various reasons this was unsatisfactory.

It was difficult and time consuming to go through the data.

Analysis by manual techniques did not always result in correct answers.

With the large volume of data many important features could be missed.

Storage of the information was expensive.

To solve many of these problems and provide a system that would allow easy and precise analysis, a PC based software package was developed. By using the then new optical disk technology, the processed data from a 300km line section would fit on one cartridge. Although streaming tape could be used it would not allow random access of the data points. Access times for the optical drive are only slightly slower than the hard disc found on many PC's. The problem with optical drives is there incompatibility and initial expense, a more universal data medium was needed that could still offer the storage size and accessibility of the optical drives. The CD ROM which is now widely available, met all the necessary data medium requirements not only for storing and analysing data but also as a means for presenting data to the client.

By automating many of the search functions, the software allows rapid screening or if desired a feature by feature step through the line. Calculation of effective dent height are made uniformly and consequently are not subject to interpretation error.

All the data is contained on the CD ROM, and any point can be called up and viewed. By interfacing with any of a number of hard copy devices, prints in colour or black can be produced as desired.

DATABASE AND DISPLAY SYSTEM

There are two database systems used in the processing and reporting of the **GEOPIG** data. The firs **GEODISPLAY**, is used initially to check the raw data for any glitches and to confirm that the raw data is good enough for processing. **GEODISPLAY** can display up to ten different sets of data consecutively, the data is chosen from a list of raw and processed databases, the following is an example of some of the available databases.

Raw Databases	**Processed Databases**
raw inertial	plan (consists of plan view, profile, horizontal and vertical percentage strain, and curvature)
raw weld	diadxdz-pcres (this gives the position of the **GEOPIG** relative to the pipewall)
raw odometer	weld (processed welds)
raw caliper	proc-atod (pressure and temperature)

Once the data is processed **GEODISPLAY** is used to find any anomilies. The data sets can be mixed and matched to allow the user to confirm features in the line , for instance, welds , which will not only have registered on the weld detects, but also in the gyro and accelerometer data. In this way by comparing data sets one can be confident of the information reported by the **GEOPIG**'s sensors. Examples of **GEODISPLAY** are given in Figs 8-10. **GEODISPLAY** is also capable of displaying two individual runs for the same pipeline and so this ability is used for change monitoring , by displaying year on year runs. This feature is especially crucial when monitoring a pipeline for fitness for purpose.

GEODENT was developed to help the user to display and analyse the processed data from Nowsco Pipeline Services intelligent pig, concerning suspected dents and their location. **GEODENT** provides two displays on a PC screen which can be selected by the user to display the data in a number of different ways.

Main Display

Thermal display (colour map of out of roundness)
Round section display (a long stack of pipe section)
Contour display

Detail Display

Thermal display
Cross sectioal display
Profile display
Round section display
Contour display
Gyro output and diameter
C and P section
Three dimensional view
Pig deflection
Pipeline path plan
pipeline path elevation

The database also supports a "dent mode" in which the user can specify the radial and linear size limits which define a wall feature as a dent, and utilise this mode to fast screen the data for reportable anomilies. Features found in this way may then be examined in detail using the detail display option. Using the various sensor outputs in conjunction with each other gives high confidence in the anomilies. For example a dent in the pipewall shows up not only as a feature in the sonar caliper output, but as a tool deflection (as the cups traverse the dent) and as an anomily in the SIMU output. Using a simple calculation based on pythagorus theorm, it is possible to compare the tool movement in the pipe to the sonar determined dent size with reasonable precision. **GEODENT** is also set up to handle more than one set of data on a pipeline, this proves invaluable when monitoring such features as preferencial weld corrosion and change monitoring. Examples of output displays from **GEODENT** are shown in figs 4-7

CONCLUSION

The use of the intelligent pig to map pipeline trajectories and geometers is now an established procedure. The data collected by the inertial tool and used with other available information has relevence at each stage of a pipelines life. The PC based software database systems make the **GEOPIG** data very accessible and therefore of infinite use for say, fitness for purpose studies, upheavel buckling studies or change monitoring of the pipeline.

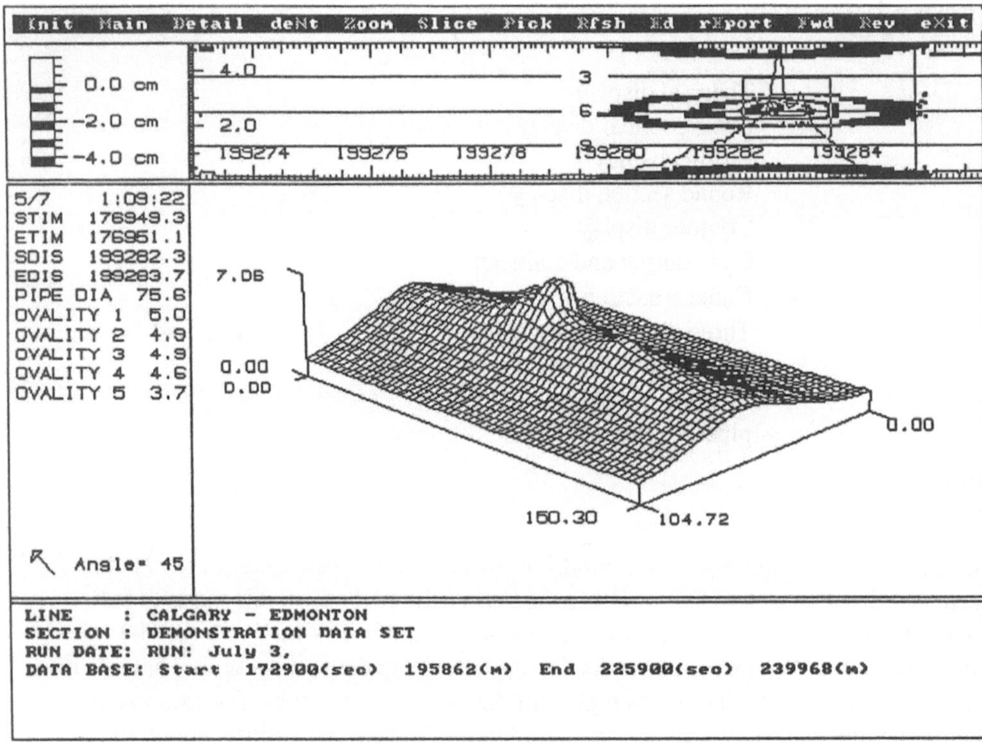

FIGURE 4 - THREE DIMENSIONAL DISPLAY OF A DENT

The gyroscopes in the Strapdown Inertial Unit are affected even by the slight movement the pig experiences in crossing a girth weld. The gyros's will be deflected by any dents.

From the geometry of the tool, it can be seen that the front cups will cross the feature some time before the sonar detect it and the rear cups will be deflected just before the feature.

By using the Gyro & Diameter Detail display and selecting a zoom window at least 2 metres before the dent, the gyro movement will verify that there is a true feature.

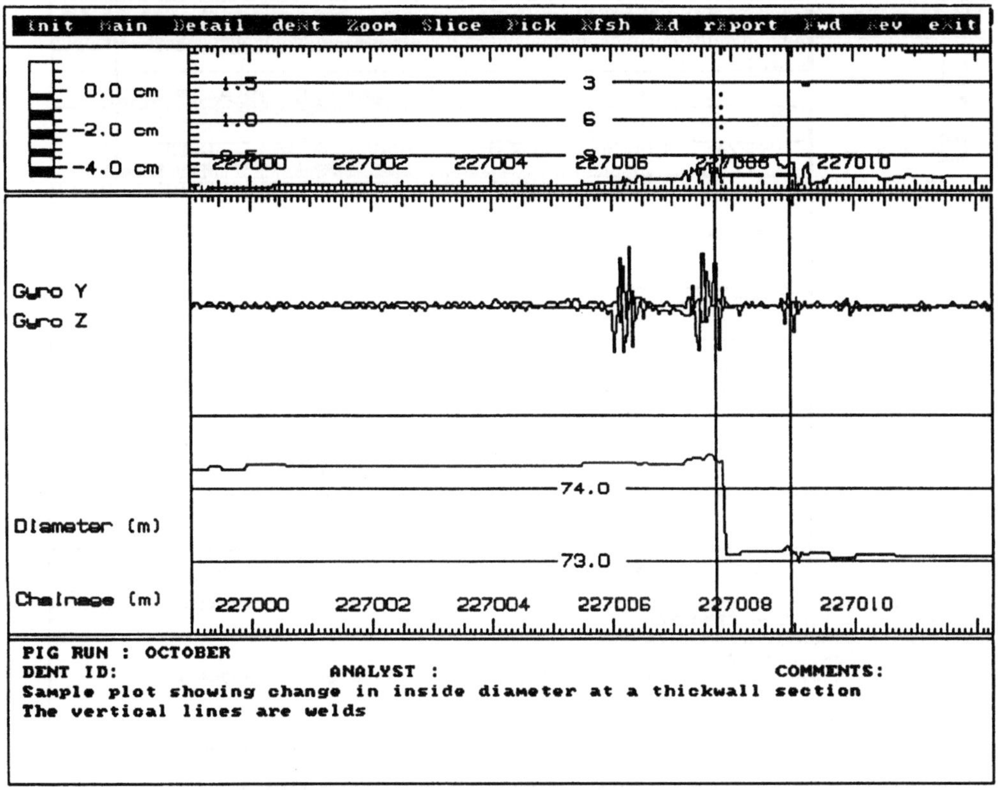

FIGURE 5 - VERIFICATION OF A DENT WITH GYROS

To check the size of the feature, the Pig Movement display is used. From geometrical considerations, the size of the dent will be three times the total movement of the front cup.

This movement will be for the NPS 30 pig 1.89 metres before the calipers measure the dent. The total movement can be calculated by taking the square root of the sum of the squares of horizontal and vertical movement.

Dents that are short in the long rack direction, (less than 0.15 m in length) will show movement as the dent falls between the pig cups. As the rear cups pass over the dent the pig will move in the opposite direction to its initial deflection. The size of the reverse move is somewhat smaller than the first movement.

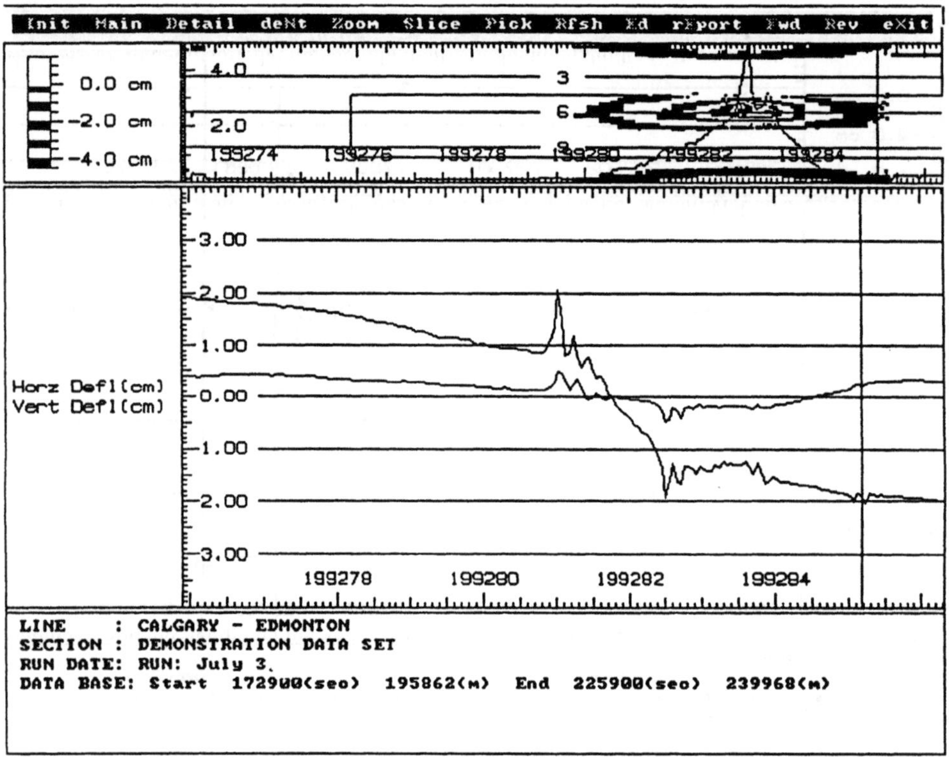

FIGURE 6 - VERIFICATION OF A DENT BY PIG MOVEMENT

The Round or Slice displays are useful in visualising the calculation that is used in determining the effective dent height. The techniques used to measure dents in the field were adapted to the measurements from the **INERTIAL GEOMETRY PIG** caliper sonar. The technique used in the field was to measure the Minimum Diameter with a pipeline caliper at the deepest part of the dent. The ovality was measured by taking the caliper reading at right angles to the minimum diameter and deducting the nominal diameter. The effective dent height was then determined by taking the Minimum Diameter from the Nominal Diameter less half the ovality. This Calculation is done automatically in the **GEODENT** Program and shows on the left box of the displays as ovality 1 through 5. The values shown as ovality is effectively the dent height for the five largest dent readings within the zoom box. The calculation is as follows:

a) The maximum deviation from the nominal pipe diameter is determined for any particular scanned section of the pipeline and the diameter is subtracted from the nominal diameter.

b) The diameter at 90° from the point of maximum diameter deviation is determined and the nominal diameter is subtracted from this value and divided by two.

c) The value in point b) above is subtracted from a) above and the result is the Effective Dent Height (called Ovality in the display) is shown and plotted.

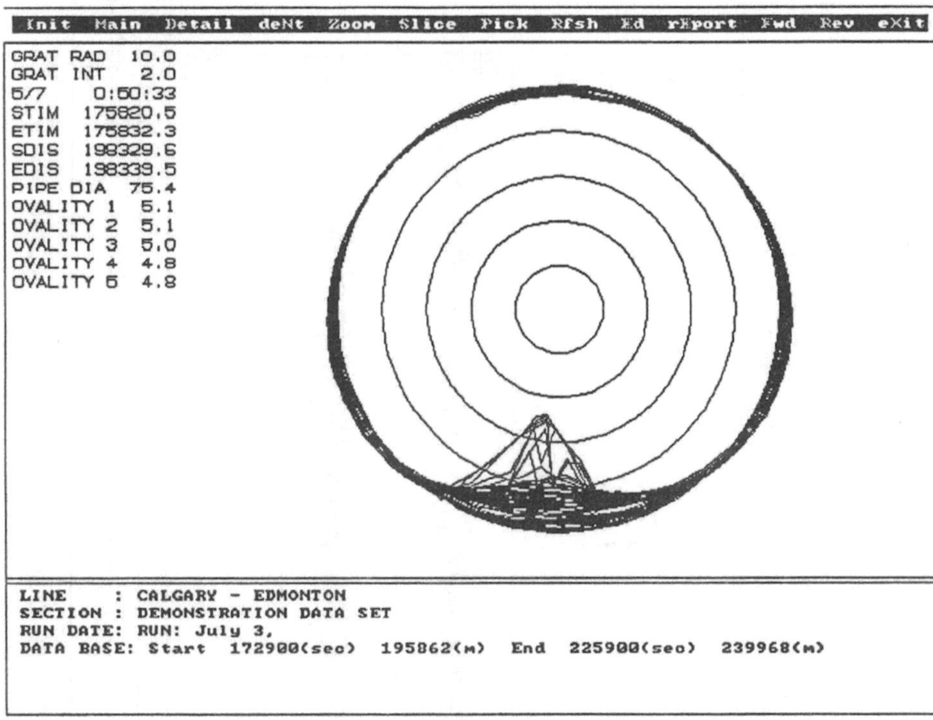

FIGURE 7 - CROSS SECTIONAL DISPLAY OF A DENT

Dent Verification

Five sites were dug up to verify the location and accuracy of the INERTIAL GEOMETRY PIG measurements. Although there was a month time lag between the measurements and the overburden had been replaced when the initial measurements were taken, all five dents compared within 1.5 millimetres (or 60 thousandths of an inch). The ovality had increased in two of the cases from the dig to the internal measurement as might be expected. Table 1 compares the results of the test dig program with the INERTIAL GEOMETRY PIG measurements.

Clock	Minimum Diameter (cm)		Effective Dent (cm)	
Position	**Test Dig**	**GEOPIG**	**Test Dig**	**GEOPIG**
7 O'Cl	74.00	74.15	2.20	1.90
7 O'Cl	75.10	74.95	1.10	1.00
6 O'Cl	74.60	74.50	1.30	1.40
6 O'Cl	73.50	73.35	2.10	2.20
6 O'Cl	74.60	74.70	1.60	1.30

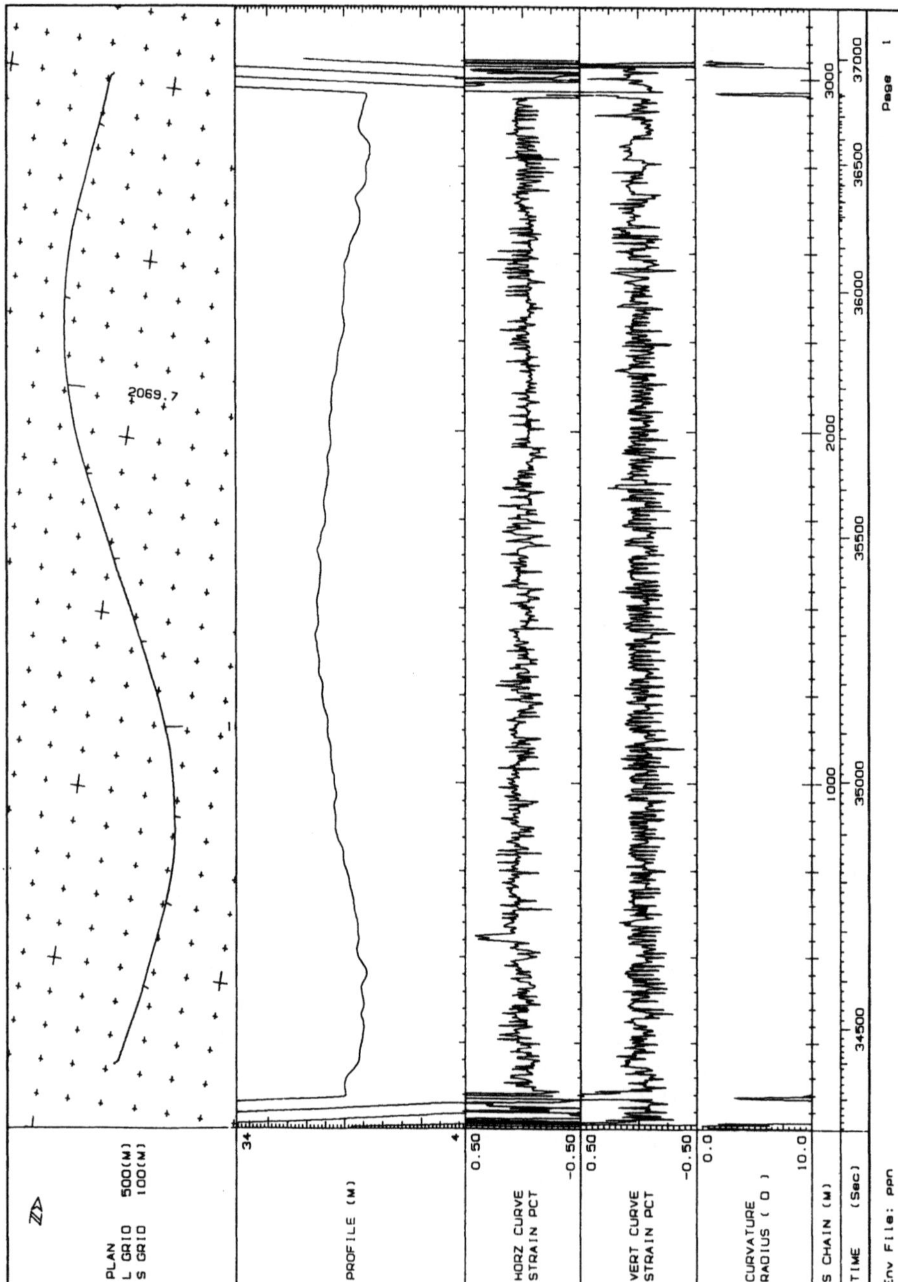

FIGURE 8 - PLAN, PROFILE AND CURVATURE STRAINS OF 3KM PIPELINE

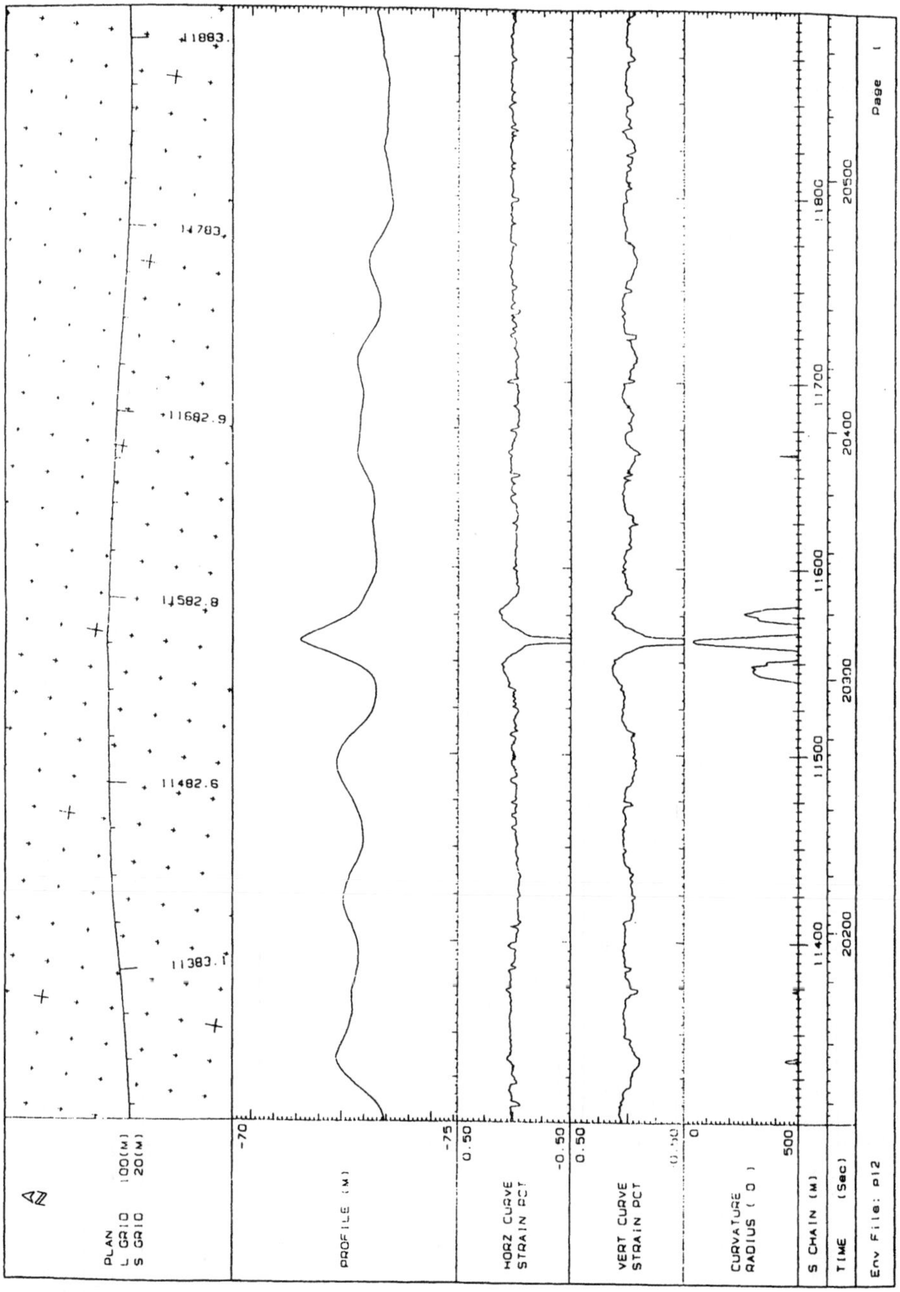

FIGURE 9 - PLAN, PROFILE AND CURVATURE STRAIN OF UPHEAVAL BUCKLE

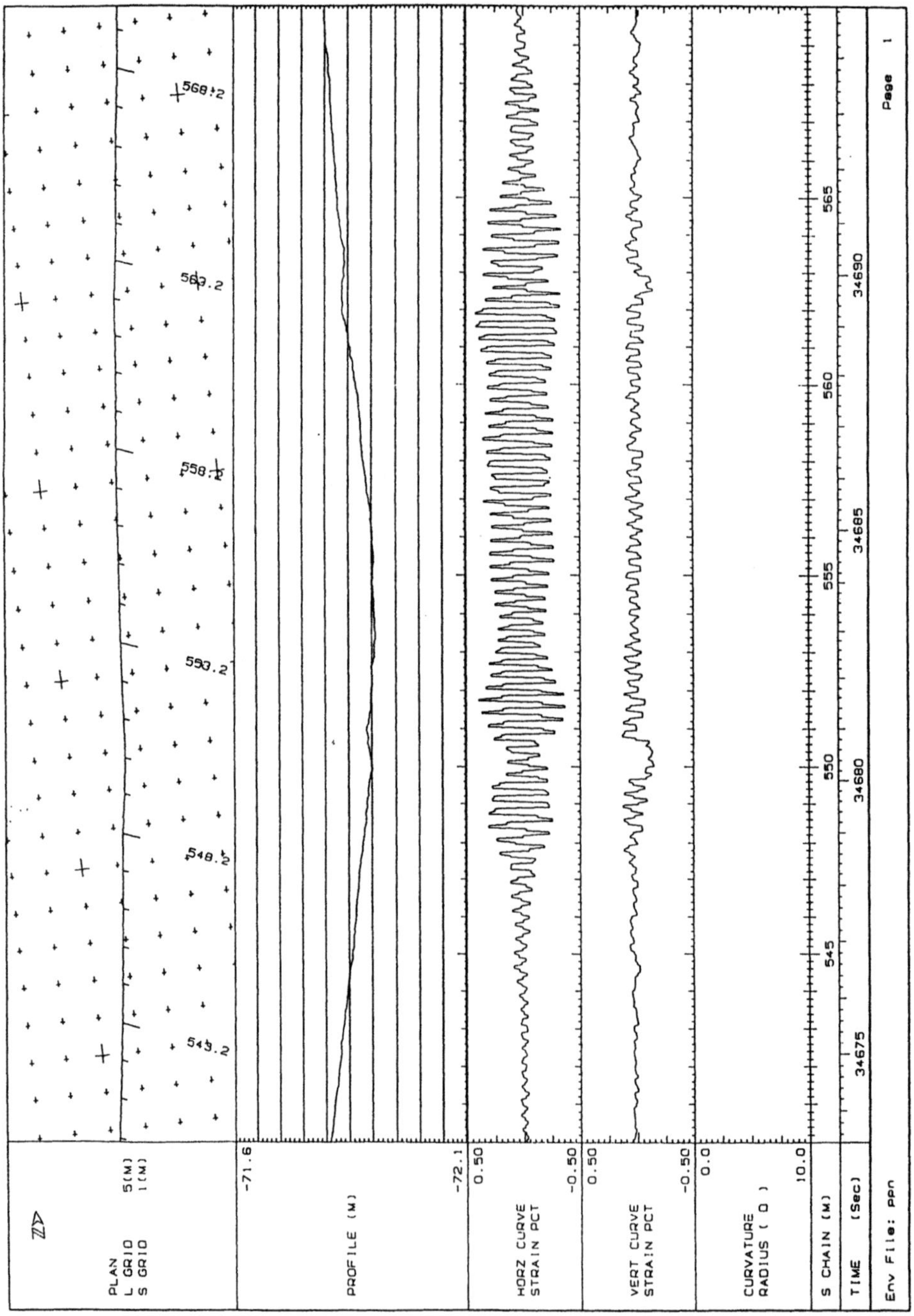

FIGURE 10 - PLAN, PROFILE AND CURVATURE STRAIN OF FREESPAN

REFERENCES

Anderson H.A. and Adams J.R. "pipeline deformation and geometry monitoring : applications of the inertial geometry pig"

Evett A. "The inertial geometry pig, its applications and field experiences"

ENSURING FITNESS FOR PURPOSE BY THROUGH-LIFE MONITORING OF PIPELINES

J H A BAKER
J P Kenny Caledonia Ltd
485 Union Street
Aberdeen AB1 2DB

ABSTRACT

Pipeline systems are major assets. The 3500 miles of pipelines on the UKCS transport to shore the majority of the £7 billion earned annually through the production of hydrocarbons. It is therefore important to maintain these assets Fit for Purpose. It is equally important to achieve this objective with minimum operating costs. To this end, limited preventative action is better than major repair: but preventative action can only be taken if an approach is adopted which facilitates identification, in good time, of any deterioration in condition which may impair the fitness for purpose.

This paper presents such an approach which is based upon the application of trend monitoring of the various inspection data, and discusses the potential savings to be made through optimisation of any rectification that may be required.

INTRODUCTION

Certain key parameters of pipeline operation, such as flow rate and pressure, can be remotely monitored. These, enhanced by a leak detection system, can provide the operators with rapid identification of a failure. However, while this is an obvious precautionary measure, the prudent operator of a hydrocarbon transportation pipeline in a hostile environment does not wait for a failure to discover that his pipeline is not fit for purpose. Indeed, he will wish to discover whether there is any likelihood of the pipeline ceasing to be fit for purpose long before it fails. To achieve this, it is necessary to carry out a regular inspection and monitoring programme.

However, because submarine pipelines and marine risers are inaccessible, the cost of inspection is high and, therefore, the frequency of inspection is limited. In addition, since measurements must be made using remote systems, the accuracy and

Volume 33: Aspect '94, 195–206.

repeatability of measurements may not be high. Any fittings along a pipeline, such as SSIVs, pig receivers, tee or wye connections, may be housed under protection structures and, as a result of this even greater inaccessibility to inspection, will suffer even more from the limitations of remote inspection. Hence, it is clearly important to employ techniques which will maximise the benefit of the inspections that are carried out. This paper describes one approach and explains the benefits to be derived.

It should be noted that, unless specifically stated otherwise, the term pipeline in this paper refers to the whole pipeline system, including risers, spools, valves and fittings.

FITNESS FOR PURPOSE

The overall aim of Through-Life Monitoring is to ensure that a pipeline remains fit for purpose. Therefore, before going on to discuss the various techniques of through-life monitoring, it is necessary to consider what constitutes Fitness for Purpose of a pipeline. It is possible to define it in terms of three inter-related conditions:

- Material integrity - cracking due to corrosion or fatigue must be avoided; mechanical working of the material is not acceptable;
- Configuration - wall thickness must not be less than that required to withstand pressures and external loads, whether due to corrosion or local damage; excessive ovality is not acceptable;
- Stress/strain - result of spanning, upheaval buckling or external loads (hydrodynamic, fishing gear, etc) must not result in over-stressing or excessive strain.

At the design stage of every pipeline, it is possible to define certain criteria, drawn from the general areas described above, by which the fitness for purpose may be assessed. The inspection philosophy for the pipeline system will then be based upon these criteria.

These Fitness for Purpose Criteria may subsequently be revised during the life of the pipeline, either in the light of revisions to codes, standards or accepted practice, or following changes in the operation of the pipeline, in which case the inspection philosophy will be modified accordingly.

INSPECTIONS

In addition to unpredictable incidents which may cause damage, there are three measurable, progressive aspects of the condition of a pipeline which influence its fitness for purpose in terms of the criteria outlined above. These are:

- internal corrosion or erosion,
- external corrosion or coating loss, and
- mechanical behaviour.

Internal Corrosion

Internal corrosion takes various forms, such as generalised wall loss, localised pitting, preferential weld corrosion and cracking, while erosion generally appears as wall loss near bends or constrictions. Although generalised wall loss might be identified by external wall thickness measurement (which is discussed further below), the usual way of identifying and measuring internal corrosion is by pigging. Generalised wall loss can be detected by a calliper pig, but measurement of pitting and cracking will require intelligent pigging. Even the latter may have difficulty identifying corrosion close to a circumferential weld while, because magnetic flux and ultrasonic pigs each have different strengths and weaknesses, no one method is ideal for detecting both internal/external defects and cracking within the wall, such as hydrogen-induced cracking.

A calliper pig will record the internal diameter along the length of a pipeline. The record can be plotted against pipe joint number since circumferential welds will show up as irregularities, and any valves or tee or wye connection orifices will also show up. An inertial navigation package can also be attached to accurately determine the true path of the pig through three-dimensional space. However, the causes of any other variation in wall thickness can not be determined.

An intelligent pig, of either magnetic flux or ultrasonic type, is calibrated prior to a run by pull-through trials in which the pig is pulled at an accurate speed through a pipeline section containing a number of specimen faults (pitting, cracks, gouges, etc). The signal in response to each known fault is recorded as a means of interpreting the run data. Since the intelligent pig will also record the passing of circumferential welds, this record will enable the plotting of all identifiable defects against pipe joint number and, hence, Kilometre Post (KP), and an inertial navigation package can again be attached. In this case, the reason for changes in wall thickness (in the sense of pitting or generalised wall loss, if not the underlying cause of the problem) can be determined.

However, the analysis of data from a single pig run may indicate that some form of corrosion is taking place, but it can not quantify the rate. Also, in practice, the identification of defects is not as fool-proof as might be hoped. Therefore, as will be discussed in the next Section, a technique is required to determine repeatability and rate.

External Corrosion

A pipeline is protected from external corrosion by a coating system (to prevent the formation of an electrolytic cell) and sacrificial anodes (to corrode in preference to the pipe wall should the coating system break down). An impressed current system may sometimes be used in the vicinity of platforms to provide a cathodic potential as an alternative to anodes. All the time the coating system remains in perfect condition, no electrolytic cell can form, and the anodes will not corrode. If the coating starts to break down in places, cells will form between the breakdowns and the anodes, the anodes will start to corrode, and a potential difference will be measurable.

The usual method of monitoring external corrosion is to measure the cathodic potential. However, this really only confirms that the anodes are still 'working' since the potential will remain high until the anodes become significantly depleted. A more effective method would be to measure anode mass depletion since this will occur from the time that the anode starts to work. Once again, because it is subjective, this is not a very accurate method of measurement, and a technique is required to determine depletion rates with some confidence.

The other method for 'flagging up' the possibility of external corrosion is to visually identify depletion of the coating system. If there is abrasion to, or loss of concrete coat, there might be damage to the corrosion coat beneath.

Mechanical Behaviour

This refers to changes in the mechanical state of a pipeline that may affect its fitness for purpose. Typical of this is when scour action causes spanning which can lead to excessive stress or fatigue damage resulting from vortex-shedding-induced vibrations. In the case of piled protection structures, scour can leave these well clear of the seabed and at risk from trawl gear snagging and pull-over, while scour around gravity covers may leave them bearing on the pipeline, again causing considerable over-stressing. Similarly, seabed mobility or the erosive effect of cyclic hydrodynamic loading may unbury a pipeline or increase its exposed length, leading to hydrodynamic instability, or risk from fishing gear interactions.

The traditional approach to inspection of these aspects is observation by ROV-mounted video cameras and data logging on the surface (the video also being recorded for subsequent review when necessary). This approach has a number of limitations [1], not least of which is the difficulty of comparing video records from survey to survey, and alternative approaches include developing transverse profiles into longitudinal profiles, or utilising digital terrain modelling techniques, fed by multi-beam sensors, to generate an accurate model of the pipeline on the seabed [2]. A method is still required, however, to assess the rates of any changes in state that may be occurring.

Another mechanical change is upheaval buckling which can occur when a hot flowline is buried with inadequate overburden. This can happen gradually (creep), possibly over several years, particularly if the production method induces thermal cycling. Because upheaval buckling occurs along the shortest possible baseline, the radius of curvature tends to be small and high stress, or even plasticity, can result.

Upheaval creep can be monitored by accurate measurement of depth of burial, and comparison from survey to survey. Clearly, this will be most effective if an accurate baseline as-trenched out-of-straightness survey is recorded prior to backfill, followed by a baseline depth-of-burial survey immediately after the backfill operation (or the following year if natural backfill is to be relied upon). However, out-of-straightness surveys depend upon accurate measurement of depth and, given current limitations in this area, a method is required to investigate relative changes in vertical position and depth of burial.

In every case, whether it is the internal, external or mechanical aspect that is being investigated, it has been shown that a method is required to reduce error bands and identify overall trends in behaviour.

TREND MONITORING AND CROSS-CORRELATION

The previous section identified the three progressive aspects which may influence the fitness for purpose of a pipeline, and reviewed inspection techniques applicable to each. However, unless the interval between inspections is very long, or the influence is very aggressive, no single inspection would be expected to show that any of these (as opposed to a single incident of damage) had reduced a pipeline to unfitness. Indeed, because of limitations in positional accuracy and the repeatability of measurements, no very obvious changes may be observed from inspection to inspection.

The overall aim of a pipeline inspection is generally taken to be to demonstrate that nothing has occurred to the pipeline, since the last inspection, to render it unfit for purpose [3]. Rather like the MoT Test for motor vehicles, it ensures that the pipeline is fit at the time of the test. The contention of this paper is that it should also be expected to identify any change in condition which could, in due course, render the

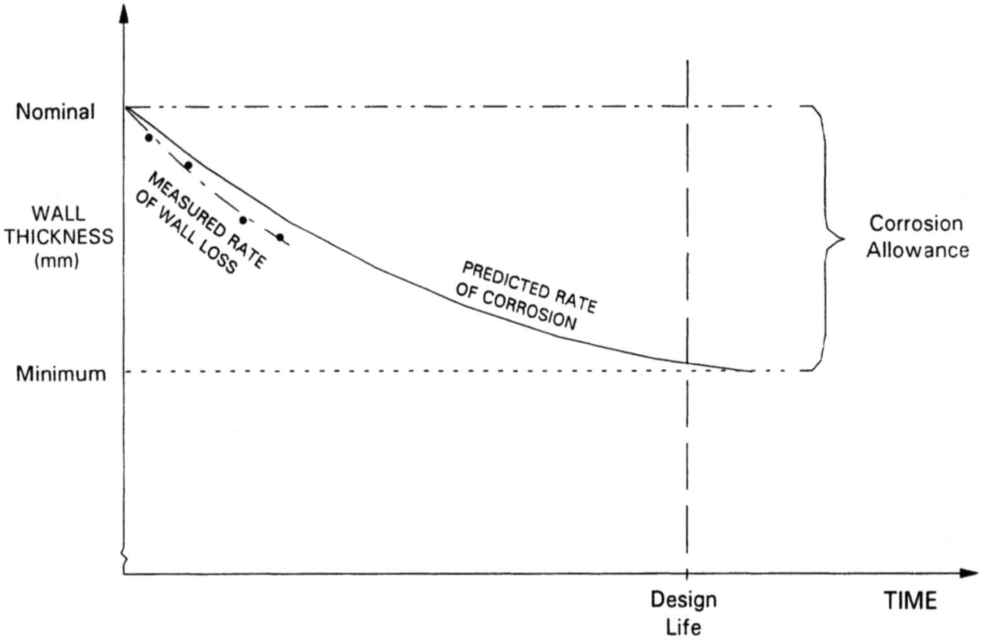

Figure 1: Example of reduction in wall thickness over time

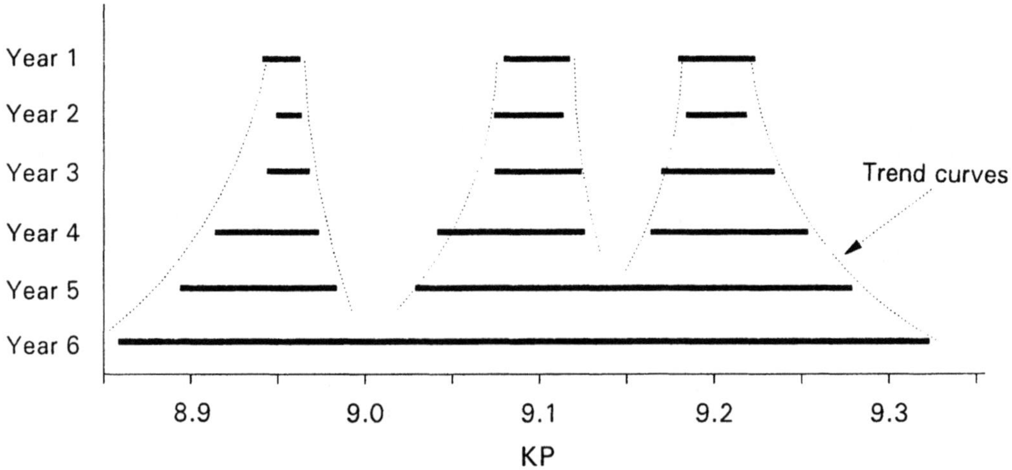

Figure 2: Example of a trend plot - increase in unburial/exposure

pipeline unfit. In this case, the primary aim of a pipeline inspection may be not so much to ensure that nothing has occurred, but rather to confirm the predictions as to the rate at which some change is expected to have occurred.

As discussed earlier, it is possible to identify certain criteria which define the limits of fitness for purpose of a pipeline system. Wall thickness is taken as an example in Figure 1. At the design stage a wall thickness is selected to ensure that the pipeline can contain the required pressures and predicted external loads. A prediction of the rate of internal corrosion is then made, and the design wall thickness is augmented by this amount so that, at the end of life, the minimum required wall will remain. In practice, the corrosion rate may vary, or the corrosion mechanism may lead to localised pitting or cracking rather than general wall loss. For this reason it is normal practice to monitor the rate of corrosion in certain key places and to extrapolate the results to the rest of the pipeline. In some cases, intelligent pigs may be run so that extrapolation is not required.

Trend Monitoring

No method of measurement is absolute, and the significance of small changes between any two inspections may not be clear. The application of Trend Monitoring can clarify the situation. Take, for instance, a pipeline where exposure is increasing: three factors may affect the reported start and end positions of exposure:

- Positional accuracy of the survey system,
- Differences in peoples' definition of 'exposure', and
- Genuine changes in exposed length.

Because of these influences, the difference between the exposed lengths reported by two surveys may not appear significant. It is only by comparing the reports over several surveys that a clear pattern emerges, as typified in Figure 2. Once clear trends can be defined for each growth position, regression techniques can be used to quantify the rates of change (Figure 3), and these rates of change can then be used to predict future change on the basis of the parameters included in the regression.

If the prediction indicates that the pipeline will become unstable in a matter of months, then remedial action must be taken. Hopefully this will not be the case, and it will be sufficient in the shorter term to monitor the changes, and to confirm the predictions. Having done so, it will be possible to determine in good time whether remedial action will eventually be required and, if so, to optimise the workscope and take advantage of long lead times in procurement.

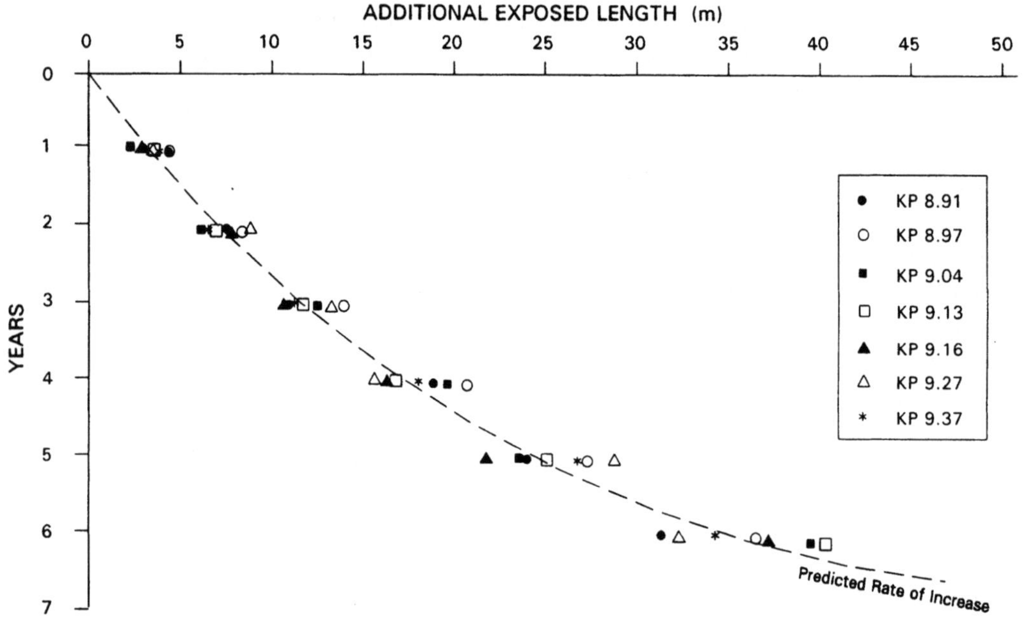

Figure 3: Example of regression - increase in unburial/exposure

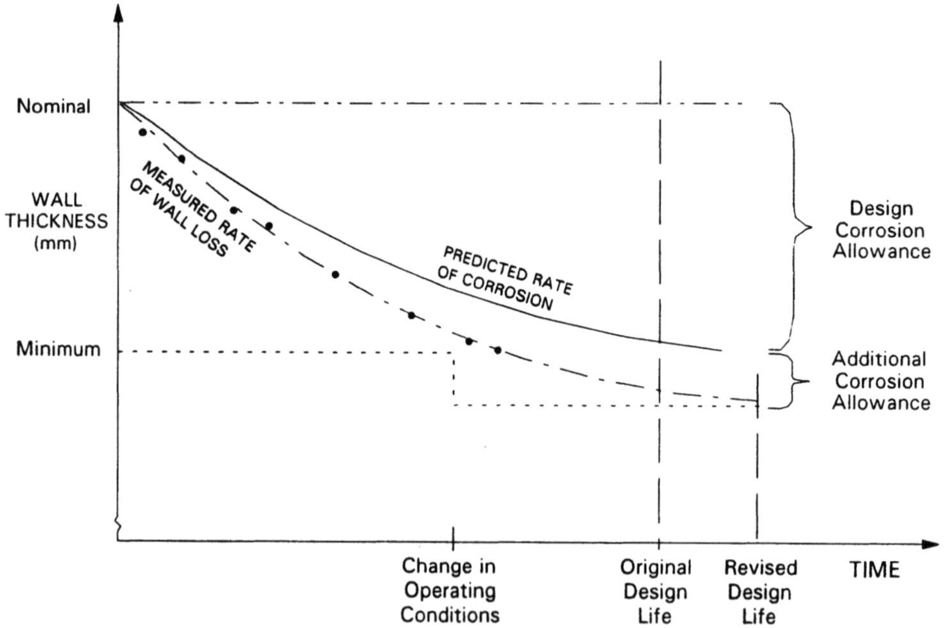

Figure 4: Influence of process change on required minimum wall thickness

Similar techniques can be applied to upheaval buckling creep. In this case, a pipetracker or sub-bottom profiler must be used to measure the depth of burial, and Trend Monitoring techniques can then be used to identify changes from inspection to inspection from what tends to be fairly noisy data. If creep is detected, additional overburden can be installed before the pipeline becomes exposed or over-stressed.

The technique can also be applied to scour-induced spans. Once scour is initiated, a span will increase in length laterally and will deflect at mid-span until it touches down [1]. This lateral growth of a span has been modelled and this can be used to predict the characteristic timescale for span set-down, in sandy soils, with notable accuracy [4]. Given this model, it is possible to predict whether a span will be over-stressed before it touches down, or whether unacceptable fatigue will accrue. As a result, pipeline inspections can again be used to confirm predicted growth rates, with trend monitoring being used to smooth out discrepancies in reported span lengths from inspection to inspection.

The above demonstrates how Trend Monitoring can be used to smooth out inaccuracies in inspection results, and to calculate rates of change of key parameters. Given a current datum and a rate of change, it is possible to predict when the state of the pipeline will fail a particular criterion, and no longer pass as Fit for Purpose. However, in practice, changes in operating practice may influence these parameters and, to be cost-effective, this must be recognised in the Through-Life Monitoring.

Cross-Correlation

To continue with an earlier example, the wall thickness of a pipeline is selected initially for the defined operating conditions. Trend Monitoring can then be used to predict if, and when, the wall thickness will be reduced below requisite. However, if, in the meantime, well depletion results in lower operating pressures or flow rates, then the required minimum wall thickness may be reduced also. This must be recognised in assessing fitness for purpose (as demonstrated in Figure 4). Conversely, operating practice may result in occasional surge pressures in which case the required wall thickness may be inadequate. It is clearly necessary to correlate inspection criteria with operating conditions.

The requirement for cross-correlation may be more obscure, however. Take, for instance, an older well: as it becomes depleted it becomes more sour and the water cut increases. Suppose that the bathymetry within the first two or three kilometres from the well is undulating. It is quite possible that water drop-out will collect in the depressions with accelerated corrosion occurring mainly at the underside of the pipeline. It is necessary to recognise this possibility because, if corrosion is monitored only topsides, the true implication of the increased corrosivity may be missed. Even if wall thickness measurements were made locally by divers (which is

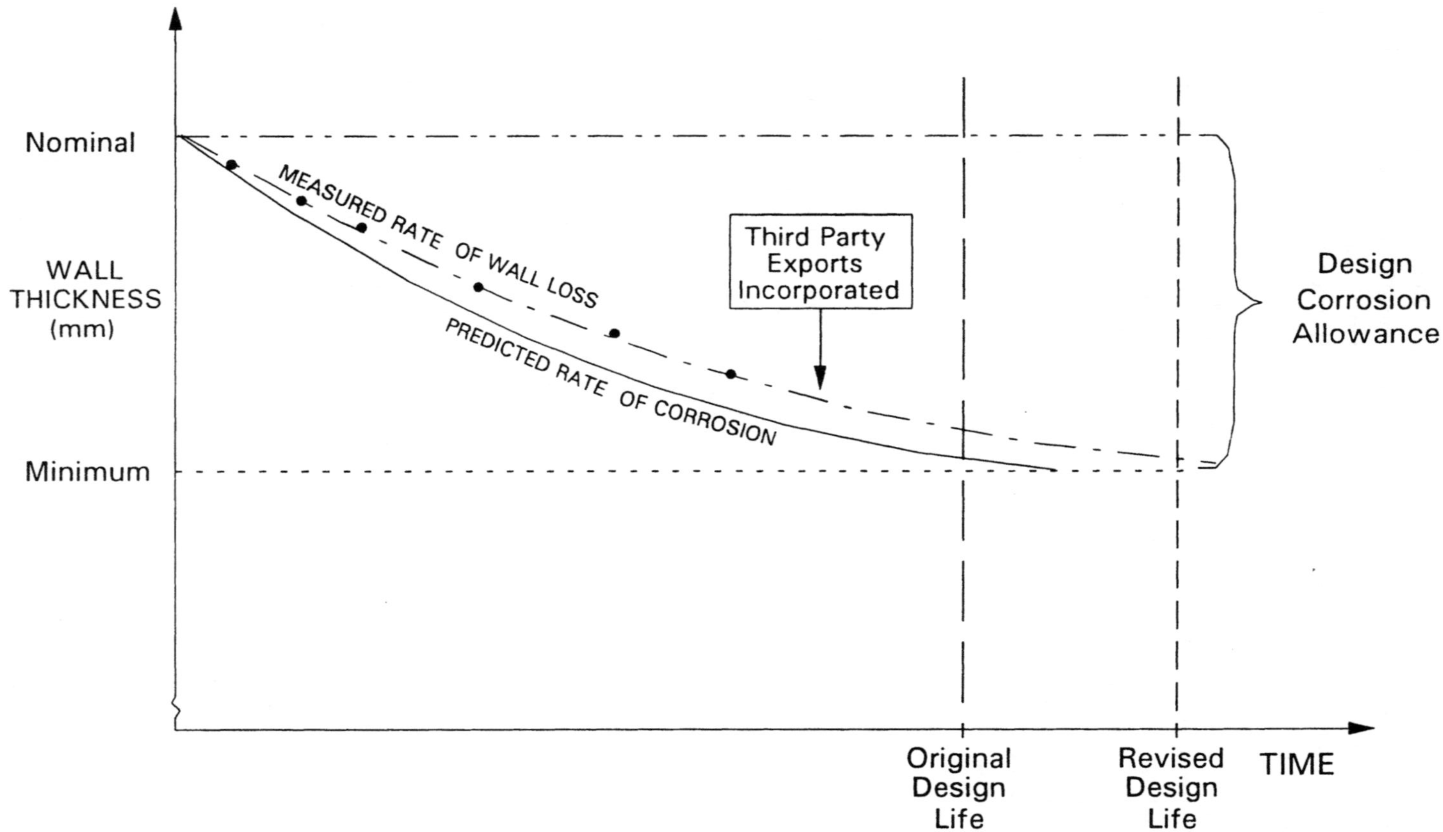

Figure 5: Benefits of through-life monitoring of loss of wall thickness

not normal practice), unless they were made at the underside of the pipeline, wall thinning would be missed.

To take the example one step further: it may be that a reduction in operating pressure is taken into account when assessing the effect of loss of wall thickness or localised pitting on fitness for purpose. It must then be borne in mind that allowable span lengths may well alter, and that the anomaly limits for spans must be revised.

The point is that inspection results can be subjected to Trend Monitoring to predict when set Fitness for Purpose Criteria may be infringed, but Cross-Correlation of data from different sources must be applied to ensure both that the original Fitness for Purpose Criteria are correct, and that unnecessary remedial action is not taken. This is of key importance in the cost-effective management of a pipeline system because, while a failure could be very costly in terms of lost production, pollution clean-up and repairs, and must therefore be unacceptable, expenditure on unnecessary remedial work will reduce the operational economic performance.

LIFE EXTENSION AND ABANDONMENT

As a pipeline draws towards the end of its original design life, two options generally arise:

- Improved reservoir performance, or opportunities for third party exports, bring a demand for extended operating life, or
- The pipeline must be abandoned.

In the first instance, this will not suddenly become apparent and the pipeline management will have recognised for some time that the design life would need to be extended. In many cases, Through-Life Monitoring of the internal corrosion would be expected to show that the operating life can be extended without further intervention (as demonstrated in Figure 5). In other cases, the original corrosion prediction curve will have been substantially correct and it will therefore be necessary to ascertain what steps can be taken to increase the life. These might include a programme to introduce inhibitors, or a planned, progressive de-rating of the line to match decreasing wall thickness (similar to the example in Figure 4). It is clear that, without the documented history which Through-Life Monitoring will have provided, it might prove very difficult to assess whether a pipeline has adequate life remaining.

If the reservoir has been expended, or the pipeline cannot be maintained in service, it will, in due course, have to be abandoned. What steps must then be taken to make the pipeline route safe will depend on the history. For instance, if it can be shown that the pipeline has remained buried throughout its life, then it will be possible to make a case for cleaning it and abandoning it in situ. On the other hand, if removal is necessary, a number of options exist. For example, reverse reeling may be considered, in which case it will be important to know whether the pipeline has the strength to be reeled. In this case, of course, the fitness for purpose criteria are somewhat different, referring mainly to whether buckling would preclude bringing the pipeline up the ramp, through the straighteners, and on to the reel. Once again, it can be seen that a well-documented history has value beyond the design life of the pipeline.

CONCLUSIONS

This paper discusses the areas from which Fitness for Purpose Criteria will be drawn as part of pipeline design, and how the pipeline inspections will be directed to monitor these. It is then demonstrated that the application of Trend Monitoring can reduce the error bands in inspection results and facilitate prediction of rates of change so that infringement of the Fitness for Purpose Criteria can be forecast well in advance. It is then shown how Cross-Correlation of data will ensure that adverse interactions are not overlooked.

In this way, Through-Life Monitoring of inspection results enables a more cost-effective approach to maintaining a pipeline system fit for purpose throughout its original design life, and beyond. It is anticipated that Through-Life Monitoring will become a standard feature of all Asset Management Programmes.

REFERENCES

[1] Baker JHA (1991) "Alternative Approaches to Pipeline Survey: The Pipeline Engineer's View", Subtech '91, Aberdeen, SUT.

[2] Baker JHA, Kirkbride P, Gallery H (1993) "Application of Digital Terrain Modelling Techniques to the Design, Installation and Through-Life Monitoring of Submarine Pipelines", Subtech '93, Aberdeen, SUT.

[3] SI (1982) No 1513, "The Submarine Pipe-lines Safety Regulations 1982", as amended by SI (1986) No 1985, "The Submarine Pipe-lines Safety (Amendment) Regulations 1986" and SI (1991) No 680, "The Submarine Pipe-lines (Inspections and Safety) (Amendment) Regulations 1991".

[4] Studies performed for BP Exploration Operating Company Ltd (not published).

TECHNOLOGICAL ADVANCES IN PIPELINE ISOLATION AND REPAIR

Dr A ALDEEN
Acurite Ltd,
Unit 7, Omni Business Centre,
Omega Park, Alton, Hants. GU34 2QD
U.K.

ABSTRACT

Pipeline isolation has been practised for many years to implement various repairs. The tools originally applied were simple and, in general, primitive.

With the emergence of the offshore industry, greater demands were imposed on pipelines. Specifications became increasingly stringent to safeguard against the arduous conditions and ensure a safe working environment. The human tragedy of Piper Alpha, vividly reported by the media, instigated an immediate review of operational safety.

The isolation tools and operational modes used in the past, are no longer acceptable as they are, at best, limited. Subsea emergency isolation valves have become essential on all major offshore pipelines. The installation of these valves has entailed decommissioning, water flooding and subsequent reinstatement activities.

Over the last decade, a vast number of pipeline isolations and intervention work has been undertaken. A variety of tools were used ranging from simple stoppers which can be manually applied through to sophisticated and tethered appliances. The latter encompass mechanical grips and hydraulically inflatable or extrusion seals. These tools were restricted to use in those areas which were readily accessible. Although umbilical hoses have allowed short distance deployment; tool retrieval has always been cumbersome.

Volume 33: Aspect '94, 207–223.

Isolation tools can be classified into three groups, past, present and future. In other words, primitive, basic and advanced. Strangely enough, the pipeline industry has been inclined to avoid modern technology. Due to the hostile nature of the North Sea, any repair tools which were instrument-dependent were considered unreliable.

With the advent of the microchip and related integrated circuit technology, the reliability of instrumentation has vastly increased. Added to this is miniaturization and ruggedness, which has enabled the development of reliable pipeline tools.

SHELL EXPRO's Sean and Leman Pipelines are good examples for assessing the advantages and limitations of past and present technologies for pipeline isolation. These 30" gas pipelines were isolated using two different techniques. Sean was isolated by means of stiff pigs, whilst remotely controlled spheres were used for Leman [1].

This paper reviews new techniques, methodology, and the experience gained during use. Emphasis has been placed on the safety aspects and the related cost savings. The special design characteristics, the reliability of communication and the enhanced safety features are highlighted.

The impact on the environment has been addressed, and in light of the future need for isolation without pollution, advanced tools will be essential.

INTRODUCTION

Pipeline maintenance and repair can be implemented safely, only under isolated conditions. To satisfy the variety of isolation requirements, simple plugging implements were initially used. With an increasing demand for improved safety it became imperative that better tools be developed. Purpose-built plugs were required to cope with higher pressures and different pipeline configurations.

Following the Piper Alpha disaster, Shell UK Exploration and Production embarked on a programme for the installation of subsea safety valves on a number of existing pipelines in the North Sea. This work was known as the "Subsea Isolation Valve (SSIV) Project". The last two pipelines to have subsea valves installed under the SSIV Project were the 30" trunklines from Leman Alpha and Sean Papa in the Southern Sector of the North Sea. These lines run to the onshore gas terminal at Bacton and are known as Sealines 1 and 3 respectively.

To implement the Health and Safety Executive's (HSE) directive for subsea isolation valves, Shell Expro considered the range of tools available for plugging these two pipelines. High pressure hydraulic plugging tools were ruled out at the outset, since they are tethered which restricted the distance they could be deployed from the platform. Freeze plugging was extensively researched by Shell Expro. However, it was considered that the freeze/thaw cycles would be lengthy and the associated costs were somewhat prohibitive. The technique for subsea operations was viewed as cumbersome and maintaining the circulation of cryogenic fluids from a diving vessel would have been weather dependent.

Due to the shallow water depth (30 - 40 metres) and the associated differential pressure requirement for the isolation, namely 3 barg, it was thought that simpler means could be used. As a result, oversized pigs and spheres were investigated to evaluate their effectiveness in pipeline isolation.

The results of a thorough programme of development and testing, undertaken by Shell Expro, showed that oversized spheres would be subjected to extensive wear during their deployment to the designated location. It was also discovered that pre-inflated spheres have a tendency to 'creep' when under pressure with the consequence of seal loss.

Stiff pigs were likewise extensively tested. The results showed the pigs to be capable of resisting movement at the specified differential pressure. Their design was refined so that they were, in effect, tailor-made for the pipeline in question - Sealine 3 (see Fig 1).

Stiff pigs, however, could not be used for the 30" Leman Alpha pipeline as it contains 1.5D bends and a riser with a restricted diameter. Furthermore, the topside pipework incorporates a 36" tee downstream of the launcher (see Fig 2). In addition, the launcher was fitted with a cup valve for loading spheres singly. Oversized spheres were not practical for the reasons given earlier. A suitable candidate was an inflatable tool, namely Acurite's remote

FIG.1 GENERAL LAYOUT OF 30" STIFF PIG,

COURTESY SHELL EXPRO

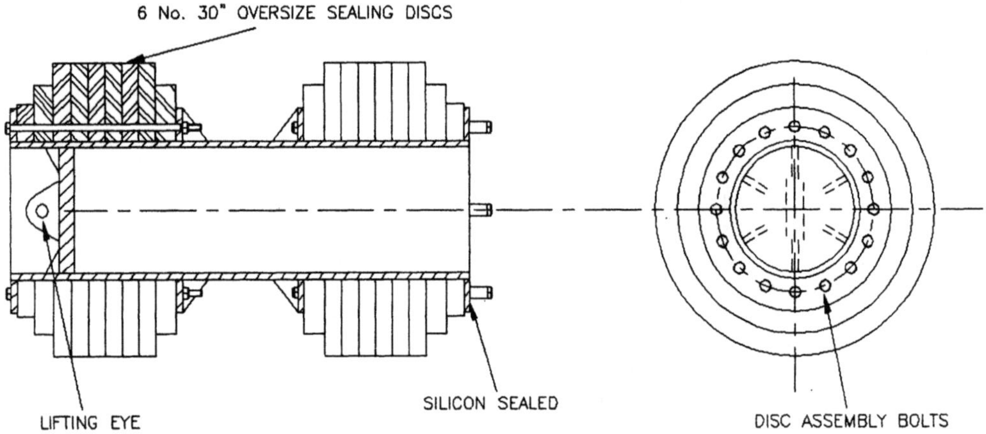

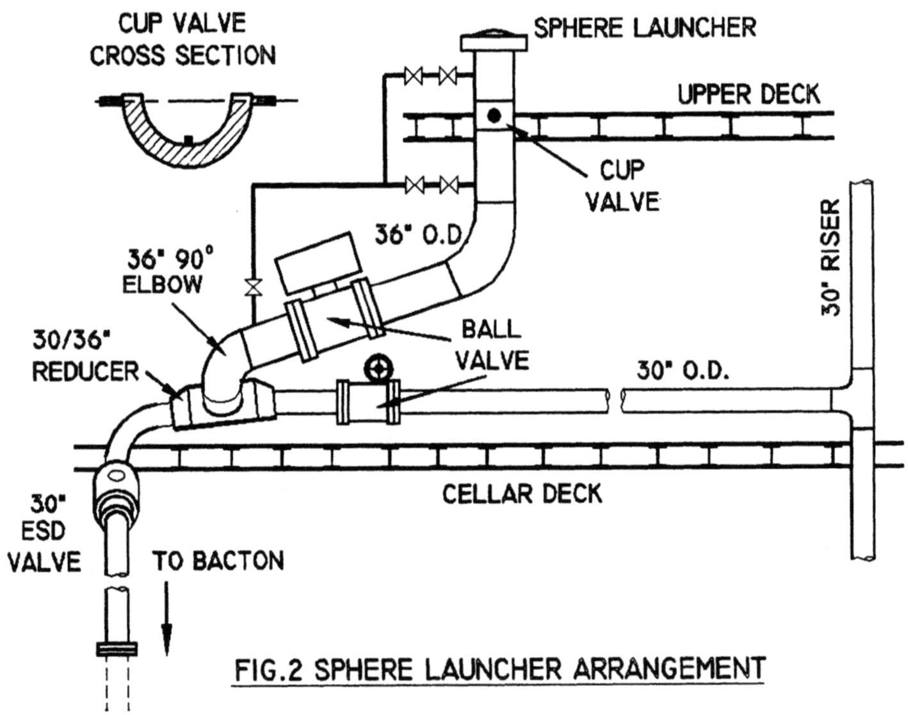

FIG.2 SPHERE LAUNCHER ARRANGEMENT

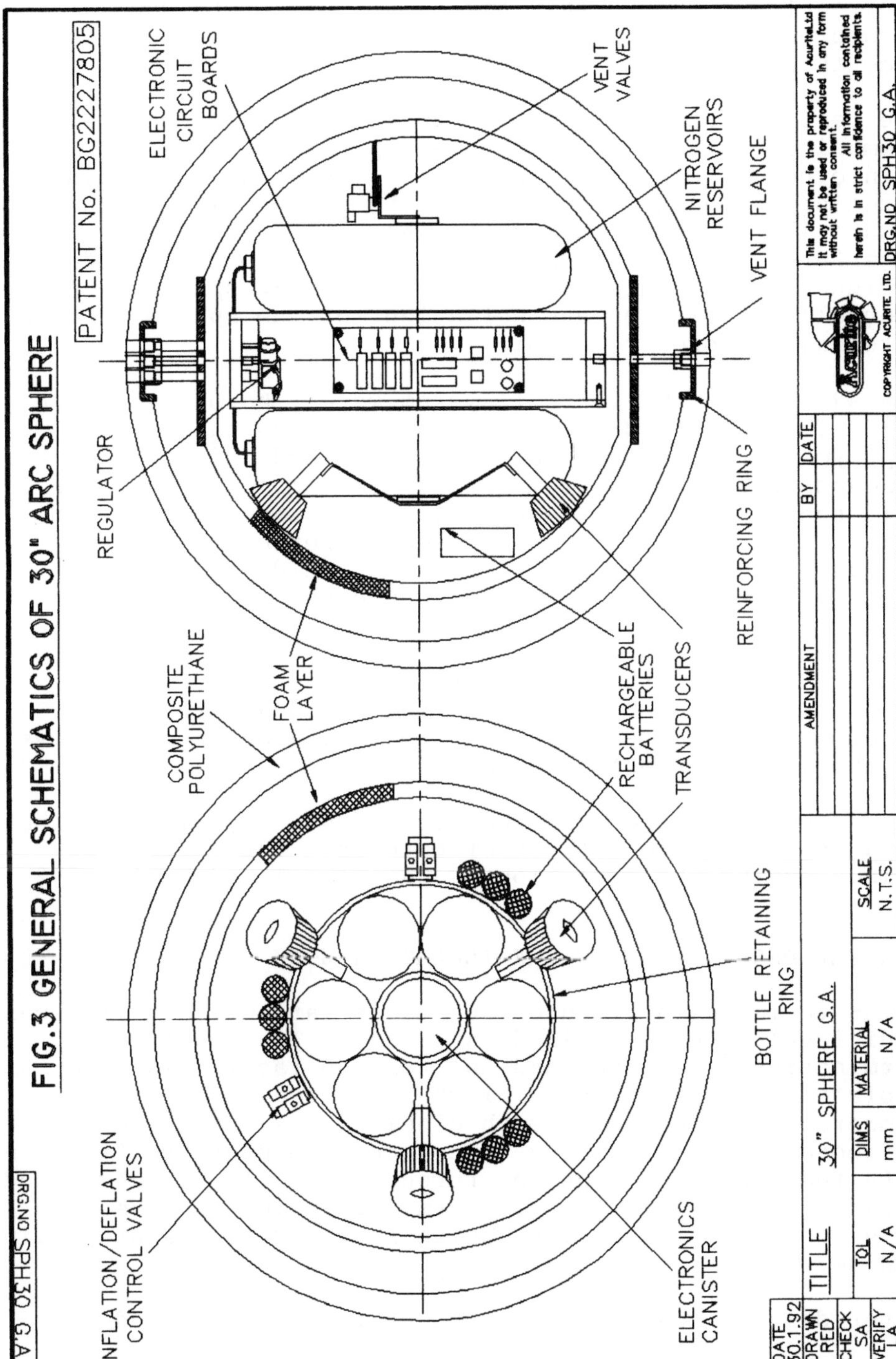

FIG.3 GENERAL SCHEMATICS OF 30" ARC SPHERE

control (ARC) sphere (see Fig 3). Shell Expro had sponsored the research and development of this tool for some three years [2].

The remote control system is capable of bi-directional communication through thick steel pipeline walls and reinforced concrete coatings. The system uses encoding, decoding and high density zero substitution of the code to ensure its reliability against the high levels of interference found offshore. The remote control system was designed to activate valves within the ARC sphere, which controlled inflation and deflation of the sphere from an internal nitrogen reservoir.

The ARC sphere and control system were rigorously tested [3]. The results proved the sphere was suitable for deployment to the designated location and inflating on commands transmitted from outside the pipe. The sphere was self-contained and self-sufficient and shown capable of maintaining seal at the specified differential pressure. It was thus decided that stiff pigs would be used to isolate Sean P and ARC spheres would be used for Leman Alpha.

Acurite's role with the SSIV Project also involved the production of the Operational Procedures and the decommissioning, isolation and recommissioning activities in the field for both pipelines.

This paper evaluates the two techniques, highlighting the advantages and limitations of both from the experience gained during the isolation of Sealines 1 and 3, in 1992.

GENERAL DESCRIPTION OF THE ISOLATION WORK

Prior to implementation of subsea isolation, each line was individually depressurised to 3 barg. The pipeline terminations at Bacton and the topside facilities were isolated using the double block and bleed principle. Temporary pipework, for each pipeline was installed offshore. These were used to connect the platform firemains with the trunkline topside, through a series of valves and metering manifolds. The platform fire pumps were used to drive the isolation train, using seawater as the propelling medium.

After driving the isolation trains to the specified

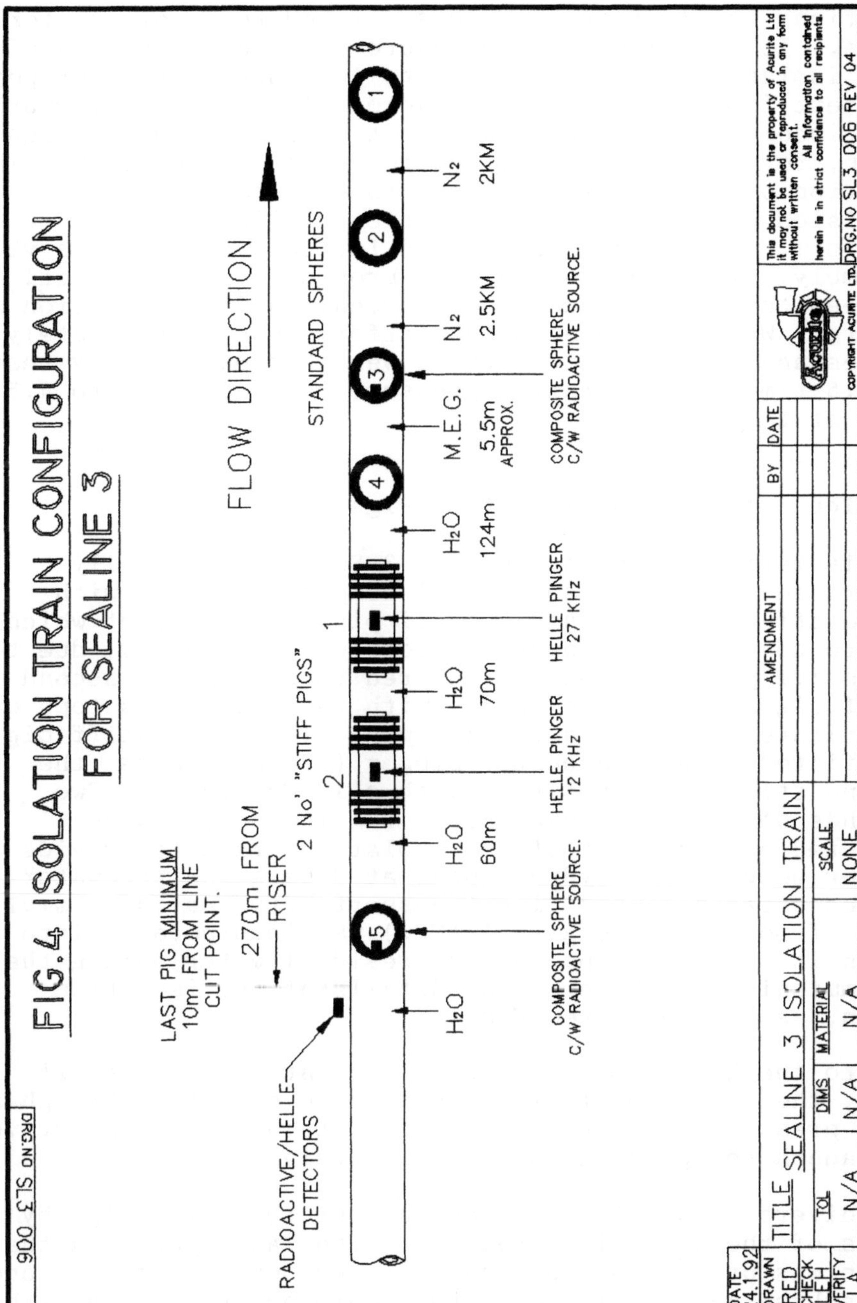
FIG.4 ISOLATION TRAIN CONFIGURATION
FOR SEALINE 3
FLOW DIRECTION
LAST PIG MINIMUM
10m FROM LINE
CUT POINT.
270m FROM
RISER
2 No' "STIFF PIGS"
STANDARD SPHERES
RADIOACTIVE/HELLE
DETECTORS
H2O
H2O 60m
H2O 70m
H2O 124m
M.E.G. 5.5m APPROX.
N2 2.5KM
N2 2KM
COMPOSITE SPHERE
C/W RADIOACTIVE SOURCE.
HELLE PINGER
12 KHz
HELLE PINGER
27 KHz
COMPOSITE SPHERE
C/W RADIOACTIVE SOURCE.
DRG.NO SL3 006
DATE 24.1.92
DRAWN RED
CHECK LEH
VERIFY IA
TITLE SEALINE 3 ISOLATION TRAIN
TOL N/A
DIMS N/A
MATERIAL N/A
SCALE NONE
AMENDMENT
BY
DATE
COPYRIGHT ACURITE LTD.
This document is the property of Acurite Ltd it may not be used or reproduced in any form without written consent.
All information contained herein is in strict confidence to all recipients.
DRG.NO SL3 006 REV 04

location, the line pressure was allowed to stabilise for 12 hours. It was important to ensure that the last tool of each isolation train travelled to just beyond the cut point furthest from the platform. Radioactive isotopes (1mCurie) were incorporated within the first and last tool of each train, for positive identification of train location. The function of the isolation train in each case was the provision of a water filled section of line near the platform, which could be cut and fitted with a subsea valve assembly. The train would also provide a positive seal between the cut zone and the hydrocarbon filled section at 3 barg. Water depth at both platforms was approximately 30 metres and the isolation trains were hydrostatically tested to 1.5 times the operating pressure which equated to 4.5 barg at LAT.

Sealine 3 Isolation with Stiff Pigs

The isolation train for the No. 3 Sealine originally comprised 3 standard spheres, 1 composite sphere with radioactive isotope, and 3 stiff pigs, the last of which was to contain a radioactive isotope. Eventually, due to excessive pressures being required to launch the second stiff pig, a composite sphere with radioactive isotope was substituted for the last stiff pig. The first stiff pig was fitted with an acoustic pinger (27kHz), whilst the second stiff pig was fitted with a 12kHz pinger. Two batches of nitrogen, each about 2km in length, were injected at the front of the isolation train to provide a buffer between the hydrocarbons at 3 barg and the water filled section. This line transports dry gas, so a small slug of monoethylene glycol (MEG) was incorporated into the train in order to "mop up" any residual water during the recommissioning phase. For details of the isolation train, as launched, see Fig 4.

Due to the high friction forces generated by the stiff pigs, the platform firewater had to be diverted through a high pressure contingency pump to obtain sufficient drive to launch the pigs and propel the train.

The divers identified the approach of the stiff pigs by means of the acoustic detectors. The last sphere in the train was propelled a distance of 10m beyond the cut zone. Two subsea radioactive detectors were positioned on either side of the pipeline. The line pressure was allowed to stabilise for a period of 12 hours. After stabilisation of line pressure, the divers confirmed by means of the radioactive detectors that the last sphere had not moved. A three hour hydrostatic leak test was carried out to validate the integrity of the isolation. The test was conducted at a pressure of 4.5 barg at LAT, corresponding to 1.5 times the subsea pressure. The test parameters were documented and recorded in terms of line pressure, topside pressure, ambient temperature and subsea temperature. Pipeline cutting and SSIV installation was subsequently implemented.

Recommissioning of the No. 3 Sealine, following completion of the SSIV installation, was accomplished by pressurising to 45 barg with gas from Bacton to overcome the static head of the riser. As all of the platform drainage and venting systems were isolated, a "well-test completion package" was hooked into the pig launcher, including surface safety valve, choke manifold, gas separator, stock tank and associated pipework and pumps. These were used to control the flow and contain any gas breakthrough.

Following pressurisation to 45 barg from Bacton and, prior to commencement of the work, all temporary pipework was leak tested then braced to prevent buffeting. Recovery of the isolation train was carried out by controlling the back flow via the choke manifold. The operation was conducted at a slow rate to prevent transients and the line pressure was maintained at 45 barg at Bacton.

As the isolation train was pushed back to the platform, seawater was dumped overboard, contaminated MEG was separated out into chemical tanks for subsequent disposal and the nitrogen was released through a temporary vent on the top deck. Ultrasonic pig detectors were installed on top of the riser and upstream of the ESD valve, to identify the arrival of the pigs and welding bladders.

All isolation tools launched into the No. 3 Sealine were successfully recovered on the platform, together with the six welding bladders installed by the divers and the line left gas filled at a pressure of 45 barg.

Sealine 1 Isolation with ARC Spheres

The isolation train consisted of 4 standard spheres, 4 composite spheres (2 contained radioactive isotopes) and 4 remotely controlled spheres. Two batches of nitrogen, each

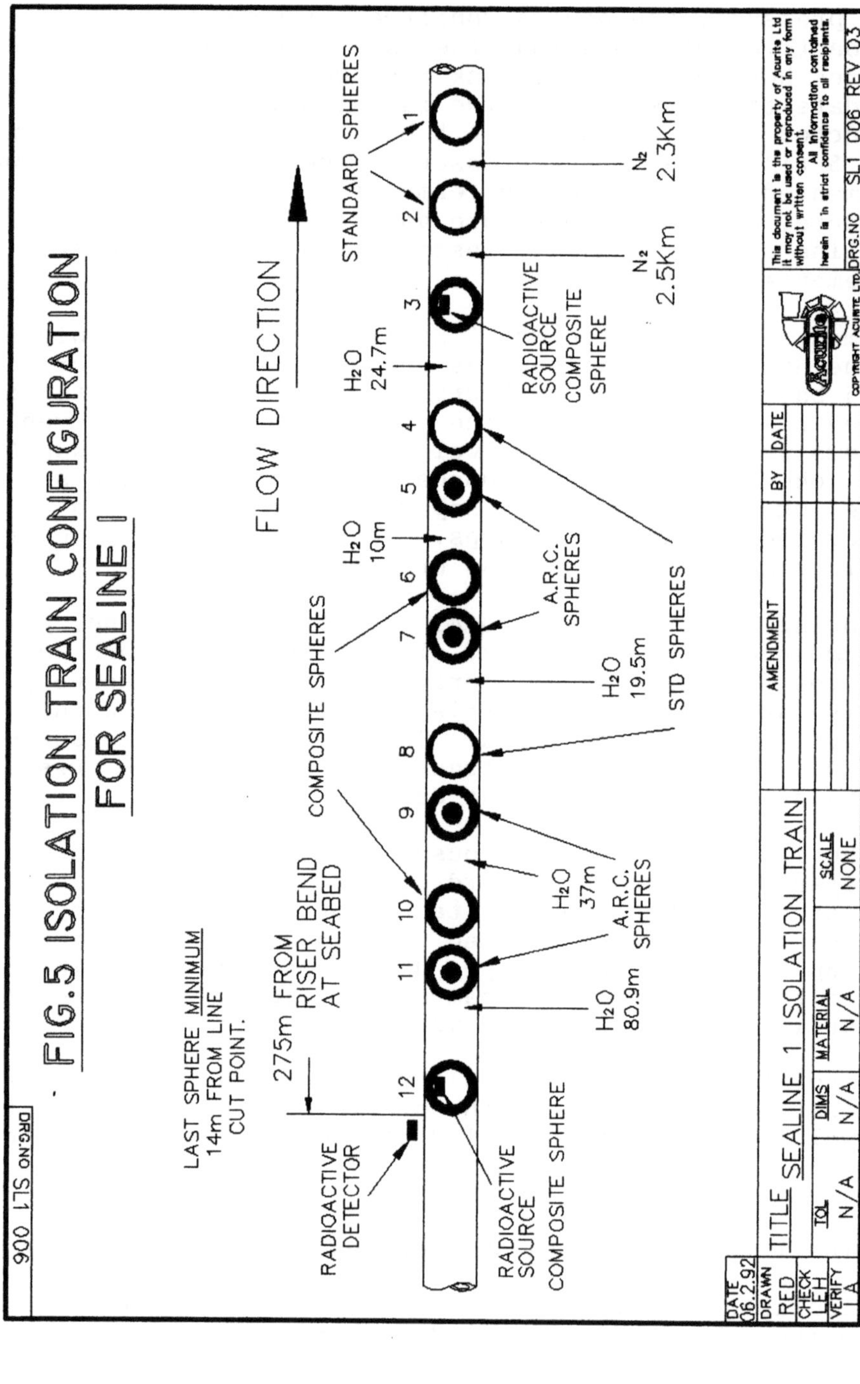

FIG.5 ISOLATION TRAIN CONFIGURATION FOR SEALINE I

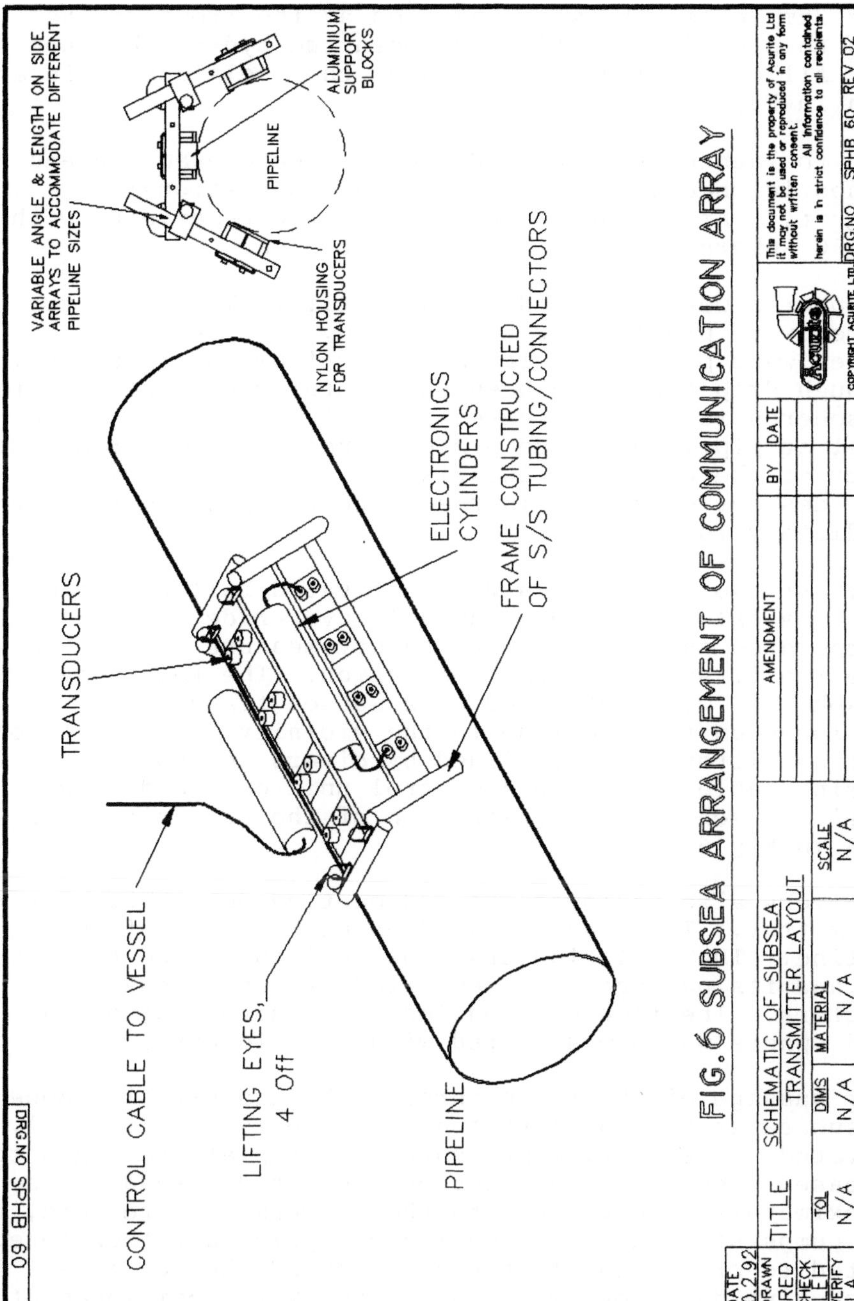

FIG.6 SUBSEA ARRANGEMENT OF COMMUNICATION ARRAY

about 2km in length, were injected at the front of the train to provide a "buffer" between the hydrocarbon and the water filled sections. Details of the isolation train are shown in Fig 5.

Prior to launch, the four ARC spheres were examined and diagnostic checks conducted to ensure operational integrity. After successful testing the reservoirs within the ARC spheres were filled with nitrogen.

The spheres were launched one at a time. The first two spheres were propelled with nitrogen. The remaining spheres were propelled with seawater from the firemains. Although drop tests had been successfully conducted on the ARC spheres it was considered prudent, since they would travel at pipeline ID, to protect these spheres from sudden impact at the bottom of the riser. In order to retard their descent down the riser a standard sphere was launched immediately in front.

The first two ARC spheres were activated following a one hour time-delay. The programmable over-ride was incorporated as an additional safety measure since this was to be the first subsea implementation. The remaining two ARC spheres were activated by a subsea transmitter positioned on the pipeline in the vicinity of the cut zone. The transmitter was positioned by divers and connected, via a marine cable (130m) to a control unit on the diving vessel. A schematic illustration of the transmitter is shown in Fig 6.

The divers also deployed two radioactive detectors near the transmitter. These were placed on either side of the pipeline. The isolation train was driven beyond the cut zone and its passage was identified by the subsea detectors. The flow was reduced after detection of the third sphere which was fitted with a radioactive source.

After passage of the train through the transmission zone of the subsea array, the last two ARC spheres were remotely activated. The driving pressure on the platform began to increase. The pressure recorder indicated a sudden rise of approximately 3 barg. Due to the increase in pressure, the fire pumps stalled. The contingency pump was, therefore, started to propel the last sphere beyond the cut zone. The radioactive detectors were subsequently re-positioned at 5 and 10 meters from the cut zone.

As the last sphere of the train passed the radioactive detectors, the flow was stopped. The final sphere of the train was positioned at a distance of 14.2 m from the weld location. The divers marked the position of the sphere, to ascertain any movement of the train. The ESD valve on the platform was closed and the line isolated for a 12-hour pressure stabilisation. The isolation train was successfully tested for 3 hours to 4.5 barg at LAT. The topside pipework was vented down to atmospheric pressure and isolated to enable the divers to commence installation of SSIV.

Following completion of the SSIV installation, four of the eight welding bladders installed by the divers, were discharged subsea. It was originally anticipated that recommissioning would involve driving the isolation train back as far as the platform ESDV for seawater discharge. During this operation, they would pass the subsea array and be deflated. Due to operational priorities on the platform, it was decided to drive the entire train, together with the four remaining welding bladders, to Bacton for recovery, after demobilisation of the DSV.

All spheres in the train were successfully retrieved and safely deflated. The work was carried out in accordance with the project schedule without any incidents. Upon examination, the spheres were found to be in good condition, considering that they had travelled over 50km fully inflated [2].

ADVANTAGES AND LIMITATIONS

The isolation and recommissioning of Sealines 1 and 3 were successfully completed.

The isolation train for Sealine 3 was curtailed, in that, instead of three stiff pigs, only two were used. The reason for this was that the pressure required to drive the 2-pig train was somewhat excessive (12 - 16 barg). The train's propulsion was associated with pressure transients in the order of 20 to 30 barg. As a result, the third stiff pig was excluded. The combination of two stiff pigs together with composite spheres was adequate for isolation.

The main difficulty encountered was the erratic movement of the isolation train. This could have been due to the discs of the stiff pigs being oversized; given the minor variations in ID in some of the topside pipework. For future applications, it would be worth optimizing the hardness of the polyurethane discs to make allowance for pipeline diameter restrictions. This would possibly make pig loading easier, since for this operation, an hydraulic ram was necessary for pig insertion. It was also possible that pressure transients were due to the limited capacity of the high pressure pump. A larger pump would have been preferable to maintain flow and reduce pressure transients. However, in this instance, the size of the pump was dictated by load limitations on the platform and restricted access to the work site. The length of the water slugs could have been reduced. This would have made train retrieval easier to control and would have limited the volume of discharge.

The acoustic pingers proved to be ineffective for pig location. Whilst the acoustic signals could be picked up; it was not possible to identify the pig's position. The 12 kHz frequency was unsuitable, as the signal could be heard coming from all directions. On the other hand, the radioactive system proved very useful for subsea pig detection, but it was extremely costly. Alternative means should perhaps be considered for future work.

On recovery, the stiff pigs were found to have sustained some damage, particularly to the disc assemblies. The rear flange of the first pig was slightly deformed and the discs were somewhat dislocated. There was also some general wear and tear on all disc assemblies. The polyurethane discs of the second pig were distorted and worn, but otherwise the pig was in reasonable condition. It should also be noted that the welding bladders were difficult to retrieve due to their elliptical shape, as they tended to become wedged in the receiver.

Although the stiff pig is a simple and robust tool, without the use of elaborate instrumentation, the operation would have been extremely haphazard. The operation was carefully planned giving special attention to safety. The operational procedures addressed in detail all necessary precautions to preserve the integrity of the pipeline and avoid pollution. The range of instruments used, played a vital part in the monitoring and control of the operation and in ensuring safety.

Stiff pigs can be effective. They are simple and cheap to manufacture. However, it is necessary for the pigs to be driven with liquid rather than gas to avoid transient movements and also reduce disc wear. The latter is a severe limitation to distance of deployment. Excessive buffeting could impair pipeline integrity and may even be dangerous.

Stiff pigs are easy to control when driven with liquid, so in oil lines they can be readily used without difficulty. Recovery in gas lines, however, can be awkward since their movement cannot be easily controlled. At the discharge port, a gas/liquid separator and a choke manifold may be necessary to control gas bypass and so maintain operational safety. The ancillary facilities and the size of plant required to drive stiff pigs substantially increases the operational costs. In other words, the overall cost and the time involved for isolation and subsequent recovery is greater than would perhaps be expected given the low initial cost of the pigs. In retrospect, it may have been preferable to have used ARC spheres rather than stiff pigs to perform this isolation.

ARC spheres were simple to deploy and operate. The isolation was accomplished in a short time without difficulty. It is evident that the total number of spheres in the train and also the number of ARC spheres could have been reduced. This obviously, would simplify the operation, curtailing both time and cost. It would have been advantageous if the ARC spheres had been deflated prior to train retrieval; their recovery would have been easier. On this occasion, however, a diving vessel would have been required in order to deactivate the spheres. On the other hand, sphere deflation could have been implemented immediately following SSIV installation, before the diving vessel left the site. In this instance, the work priorities on the platform dictated the mode of operation.

The advantages of the ARC sphere are outlined as follows:

1. Self-sufficient, self-regulating and remotely controlled.
2. Untethered and deployment distance is unlimited.
3. Sphere wear is negligible since it is deployed uninflated.
4. Versatile in traversing restrictions, bends and tees.
5. Applicable in gas and liquid pipelines.
6. Simple to deploy and operate.
7. Robust and reliable.
8. Reusable since it can refurbished.

By virtue of its shape, the ARC sphere is extremely versatile, but its pressure holding capability is only moderate, namely 3 to 4 barg. In combination, however, multiple spheres can withstand higher pressure differentials. This though will increase the cost of the operation. The cost could be reduced by mass production, if spheres were stocked ready for emergencies.

CONCLUSIONS

Two isolation techniques have been reviewed. These were based on the experience gained on the SSIV Project. In both cases the technique involved could be considered to be simple and to some extent conventional. Standard materials and manufacturing processes were used. The stiff pig can be regarded as a primitive concept involving oversized disc assemblies to resist movement and exert extra frictional resistance. The novel distinction with regard to the ARC sphere is that it is self-regulating, self-sufficient and inflatable by remote control.

The project objectives were accomplished in total safety and without difficulty. Flooding the entire pipelines with water was avoided, which is of great significance. Pollution was reduced to a bare minimum due to restricted water discharge. Only a limited quantity of gas was vented during pipeline depressurisation. Minimal disruption was caused to platform operations. Operational shutdown was substantially reduced. As a consequence, considerable cost savings were achieved [4].

The next generation of remotely controlled pigs will be multi-sealed and equipped with telemetry systems, enabling the integrity of the isolation to be frequently monitored from a DSV. This will greatly enhance operational safety and promote operator confidence.

Acurite has lead the way in through-pipe communication and intends to remain among the leaders in this field. The challenge ahead will be the provision of advanced telemetry tools which can be accurately positioned at any point in a gas pipeline. Research and development in this direction is progressing and the first prototype ARC pig has been completed. Details of the second generation of remotely controlled plugs will be reported when all field trials have been concluded.

REFERENCES

[1] Aldeen, A. (1993) "Remote-Control Spheres", Pipeline Pigging and Integrity Monitoring Conference, Houston, Texas.

[2] "Isolation System Reduces Gas Line Downtime", *Offshore Engineer*, September 1992.

[3] Newman, W. J., Kontou, T. and Aldeen, A. (1993) "Pipeline Isolation Using Remotely-Activated Spheres", Offshore Technology Conference, Houston, Texas.

[4] Aldeen, A. (1994) "Pipeline-Isolation Techniques Without Water Flooding", Pipeline Pigging and Integrity Monitoring Conference, Houston, Texas.

ACKNOWLEDGEMENTS

Acurite Ltd wishes to thank Shell UK Exploration and Production for sponsoring the research and development of the ARC spheres.

UNDERWATER PIPELINE REPAIR IN DIFFICULT SEABED CONDITIONS: PROBLEMS AND SOLUTIONS

B A JONES
British Gas Plc
Exploration & Production
100 Thames Valley Park Drive
Reading, Berks RG6 1PT
UK

N WRIGHT
British Gas Plc
Research & Technology
Engineering Research Station
Newcastle upon Tyne NE99 1LH
UK

1.0 SUMMARY

This paper examines the application of innovative technology to significantly reduce the time for a major repair to a 36 inch diameter buried submarine pipeline.

The actual repair is achieved by cutting out the damaged section and inserting a replacement spool-piece. Such operations would be based upon hyperbaric welding which is performed in a dry habitat chamber and is a relatively controlled and straightforward operation. Therefore the research effort was directed towards saving time on the auxiliary tasks such as: seabed excavation, the use of a protective cofferdam, seabed surveying, pipeline weight coating removal and pipeline isolation.

It is anticipated that a 40% reduction in the overall repair time is possible with the application of the equipment and methods described in this paper.

2.0 INTRODUCTION

At present, British Gas operates two offshore gas fields in the territorial waters of the United Kingdom (UK), Morecambe on the West Coast and the Rough Storage Facility on the East Coast. These fields play a vital role in the supply of gas to the UK National Transmission System.

In each field the gas is transported to an onshore terminal through a single 36 inch diameter submarine pipeline. The pipelines were buried to a depth of 1.5 m and coated with a high strength concrete weight coating to reduce the chance of damage from anchors or fishing trawl boards. While these measures increase the security of the gas supply, repairing such a submarine pipeline in the hostile waters off the UK, represents a serious technological challenge. Traditionally all subsea work is performed by a few

Volume 33: Aspect '94, 225–247.

divers using hand held tools who are remotely directed by a supervisor located on a surface vessel. It was estimated that using such practices to excavate and repair the pipeline, would take over 60 days. Faced with such a potential down-time, British Gas instigated a research and development programme to improve the productivity of subsea pipeline repair techniques.

Analysing the overall operation of repairing a buried submarine pipeline, shows that the majority of the time is allocated to auxiliary tasks such as: pipeline excavation, concrete weight coating removal, pipe alignment and dewatering/drying the pipeline after the repair. The actual welding operation associated with inserting a spool pipe to replace the damaged section is performed in a hyperbaric (ie dry) habitat chamber[1,2], and is a relatively controlled and straightforward operation. Therefore, the research effort was directed towards the auxiliary tasks.

In order to perform underwater research in a cost-effective environment, the Engineering Research Station (ERS) established a Subsea Engineering Centre at Blyth[3]. This is located 15 miles north of Newcastle upon Tyne and is based upon a redundant dry dock facility. The largest dock was converted to allow a constant 10m water depth to be maintained. This facility allowed the evaluation of new subsea technologies in a realistic underwater environment which provided a cost-effective and rapid means of developing new equipment

3.0 SEABED EXCAVATION

To carry out a major repair on a buried pipeline, sufficient clearances are necessary to install equipment. This requires a 4m deep excavation. Depending upon the severity of the damage, the excavation could be up to 18m long and 6m wide to accommodate the positioning of the welding habitat. This would require the excavation of at least 450m^3 of seabed material, but because the angle of repose of soft seabed materials can vary from 20° to 5°, the amount of excavation required could increase to between 1,200m^3 and 3,500m^3 respectively.

In order to quantify the problems associated with such a large subsea excavation and examine the efficiency of various excavation devices, an artificial seabed was constructed in the Subsea Engineering Centre. The seabed was constructed over a 36 inch diameter concrete coated pipeline and measured 64m long, 18m wide and 4m deep. Over 10,000 tons of material was used to construct the seabed with half its length comprising of soft sand and gravel, representing the conditions found in the Morecambe Field; while the remainder was well compacted boulder clay of the type found in the Rough Field, Figure 1.

An extensive evaluation programme into seabed excavation was then undertaken for soft sand and gravel seabeds. Conventional water-energised jet pumps were found to be the most reliable solution, although their excavation rates vary significantly from as low as 3 up to 30 m^3 per hour. This variation in rates depends upon how effectively the suction heads are deployed. For higher excavation rates, British Gas have co-funded research into propeller driven devices (Jet Prop 25,000) which produce a down-draught of large

Fig. 1 Construction of artificial seabed – Blyth Subsea Engineering Centre

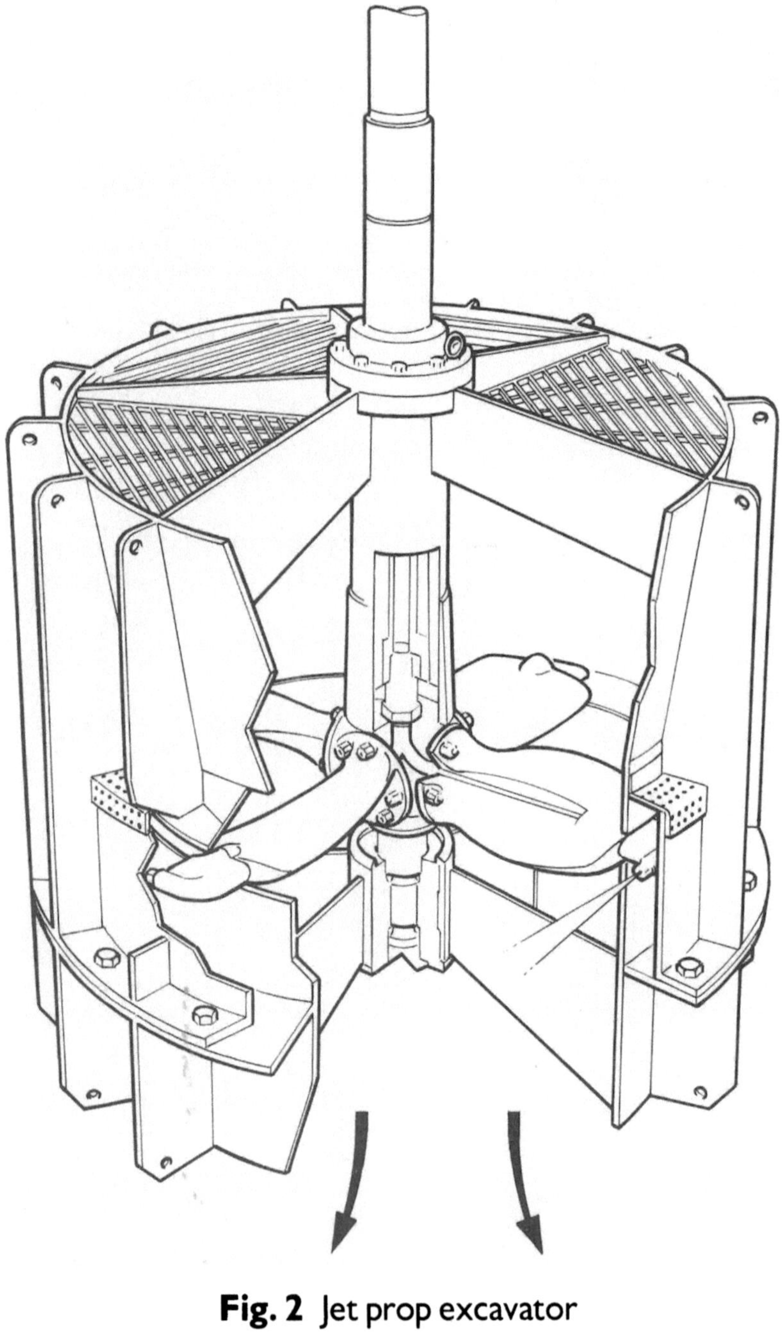

Fig. 2 Jet prop excavator

volume, low velocity water which fluidises the seabed and enables the spoil to be carried away by the prevailing sea currents, Figure 2. These devices can give excavation rates which can exceed 200m^3 per hour in soft seabeds.

SELF-BURYING COFFERDAM

A self-burying cofferdam was designed and tested to minimise the amount of spoil to be excavated and provide a safe haven for the diver in the deep excavation, Figure 3. Self-burial was achieved by diverting water from an upper ring main (built into the top of the cofferdam), into a series of jet pumps located within the walls of the cofferdam. Water was also diverted into a series of jet nozzles mounted within the lower skirt of the cofferdam to fluidise the seabed, Figure 4.

The cofferdam was successfully tested at Blyth where a self-burial rate of approximately 1m^3 per hour was achieved during underwater trials, Figure 5.

CLAY EXCAVATION

Excavating boulder clay is far more difficult. This is found in the seabed around the Rough Field with strengths of up to 250Kpa. Digging wheels with protruding teeth or picks suffer from impact from stones found within the clay and therefore are prone to poor reliability. To overcome these problems a novel clay cutting device was designed by British Gas and is based upon a 1m diameter rotating disc with strategically placed high pressure water jets to break up the clay. The whole assembly is rotated around a shaft with a built-in 8 inch jet pump to discharge the spoil, Figure 6. The clay cutter was mounted on an underwater crawler for remote excavation trials on the simulated seabed. Excavation rates of 10m^3/hr were achieved in 100Kpa clay without the aid of divers. This is ten times faster than any other system tested in similar conditions. The system has since been successfully trialled offshore, excavating the seabed for a 40 inch diameter pipeline in the North Sea.

SEABED SURVEYING

One of the main conclusions from the excavation trials was that irrespective of the method used; the control of this operation is extremely difficult because the zero visibility working conditions means the topography of the seabed is unknown during the excavation operation. In turn, this means the excavation equipment cannot be effectively deployed.

British Gas have therefore developed an advanced surveying system called SONARGRAPHICS to provide a fast-response surveying technique providing 3D plots and contour maps of the seabed. The system is based upon commercially-available sonar equipment, which provides a cross-section profile of the seabed. The resultant two-dimensional profiles are, however, difficult to interpret. Therefore British Gas has designed a system of scanning the seabed with a regular pattern of ultrasound that can be translated into a 3D picture of the seabed topography. To obtain this precise pattern of ultrasound the sonar transducer head rotates in 1.8° steps, which in turn is mounted onto an arm which is computer controlled to move in 3° steps in a plane normal to the axis of the sonar transducer, Figure 7. The whole assembly is mounted on a passive buoy tethered to the seabed by a clump weight. The movements of the buoy induced by the sea currents and wave action are measured from in-built instrumentation and gyros and this data is fed into the computer to produce an accurate seabed survey.

The raw data can be verified by the operator who can also manually remove spurious points from sound reflections from fish or umbilicals using the 'mouse' facility on the computer terminal screen. Once the operator is happy with the quality of the survey information, the data is processed by the computer to produce 3D plots of the seabed and contour maps. Figure 8 shows a 3D plot of an excavation either side of a 36 inch diameter pipeline taken in 120m of water. The associated contour maps can be interrogated by the operator to produce cross-sections of features and volume plots.

The system can survey an area of 8,000m^2 from a single position 20m above the seabed with an accuracy of ±200mm. Other applications include pipeline river crossings, harbour works, scour profiles around offshore rigs and even marine archaeology. An offshore survey company, Subsea Surveys, has been granted a licence to commercially exploit the SONARGRAPHICS technology.

4.0 CONCRETE WEIGHT COATING REMOVAL

British Gas has a high strength concrete weight coating to give the maximum protection against damage. Depending upon location, the weight coating can be up to 125mm thick with 8mm diameter circumferential and 5mm axial reinforcing bars (concrete coating details given in Appendix 1). In addition, pipelines on the Rough Field have 1m of Pri-Grip adhesive every 11m, bonding the concrete to the epoxy external coating. Using such a high integrity weight coating brings new problems to the subsea engineer when the coating has to be removed for pipeline maintenance or repair.

British Gas has evaluated numerous concrete removal techniques at the Blyth Subsea Centre. These ranged from hand held hydraulically driven circular saws, which took 18 hours to remove a 1m length of coating and damaged the pipe in the process, to water jetting with grit entrainment and shaped explosive charges. Details of various methods are given in Table 1.

British Gas' repair procedure called for a 2m length of concrete coating to be removed in less than 6 hours. This target demanded a 400% improvement to the cutting time recorded on the previous trials. Therefore British Gas decided to design and build their own system[4]. The removal method is a two-stage process in which the section of concrete to be removed is first cut in a specific pattern and the resulting panels are then prised from the pipe. Cutting is carried out by a diver-operated machine with a hydraulically driven diamond tipped circular saw. The machine is strapped to the pipe by a chain

TABLE 1. COMPARISON OF CONCRETE WEIGHT COATING REMOVAL TIMES FOR VARIOUS TECHNIQUES

Removal Technique		Cutting Time	Removal	Total Removal Time	Panel Length Removed
Cutting	Removal	(Hours)	(Hours)	(Hours)	(m)
Handheld Saw*	Hand Driven Wedge, Crow Bar & Levers*	18	6	24	1
Diamond Tipped Saw Mounted on Semi-Automatic Machine	Hydraulic Jacking	7	5	12	1
High Pressure Jetting with Grit Entrainment*	Shape Charge Explosives	14.6	1.8	16.4	1
British Gas plc Cutting Machine	British Gas plc Splitting Machine	2.7	1.1	3.8	2

(Times recorded from trials held at British Gas Subsea Engineering Test Centre, Blyth)
*These methods damaged the pipeline.

system and the cutting head can be rotated to produce either circumferential or longitudinal cuts, Figure 9A. Development work was undertaken to enable the high speed diamond tipped blades to operate with maximum power consumption and acceptable blade life. The result was the use of a constant-volume, variable-pressure hydraulic power supply and an air injection system around the blade.

Cutting is arranged to produce two semi-cylindrical shells which can then be prised away from the pipe using a second machine shown in Figure 9B.

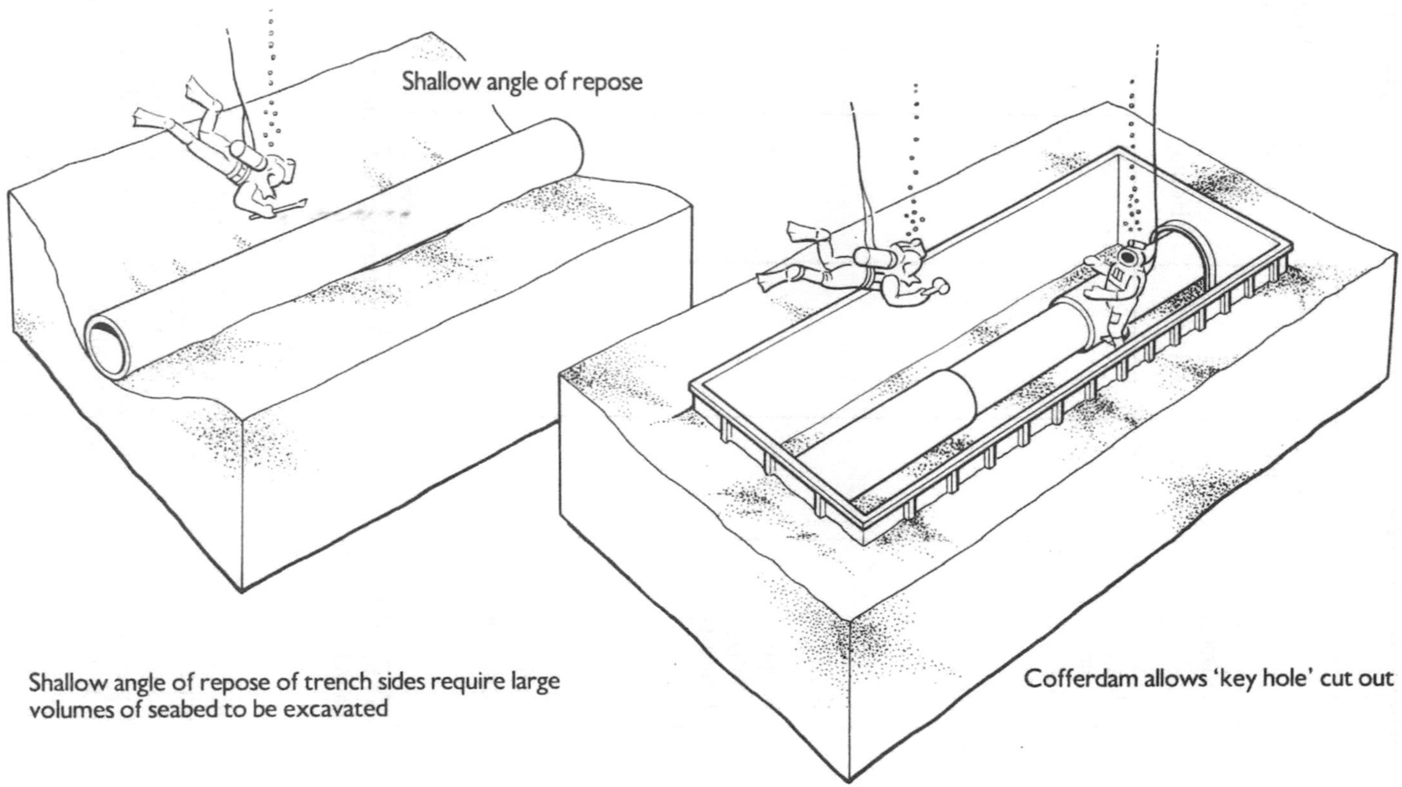

Fig. 3 Open trench and cofferdam excavation strategies

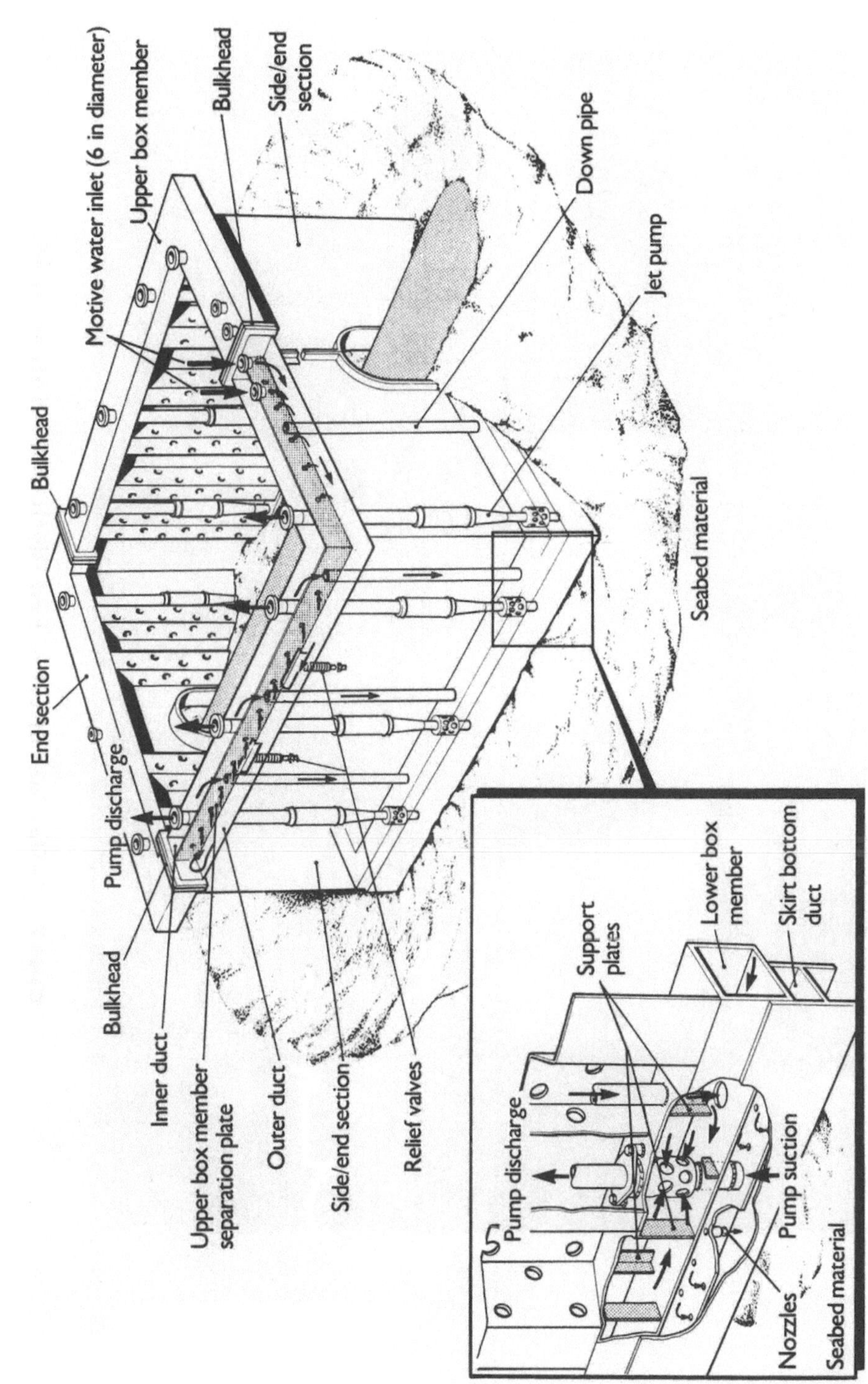

Fig. 4 Schematic view of self burial cofferdam

Fig. 5 Cofferdam lifted out of Blyth seabed after trials

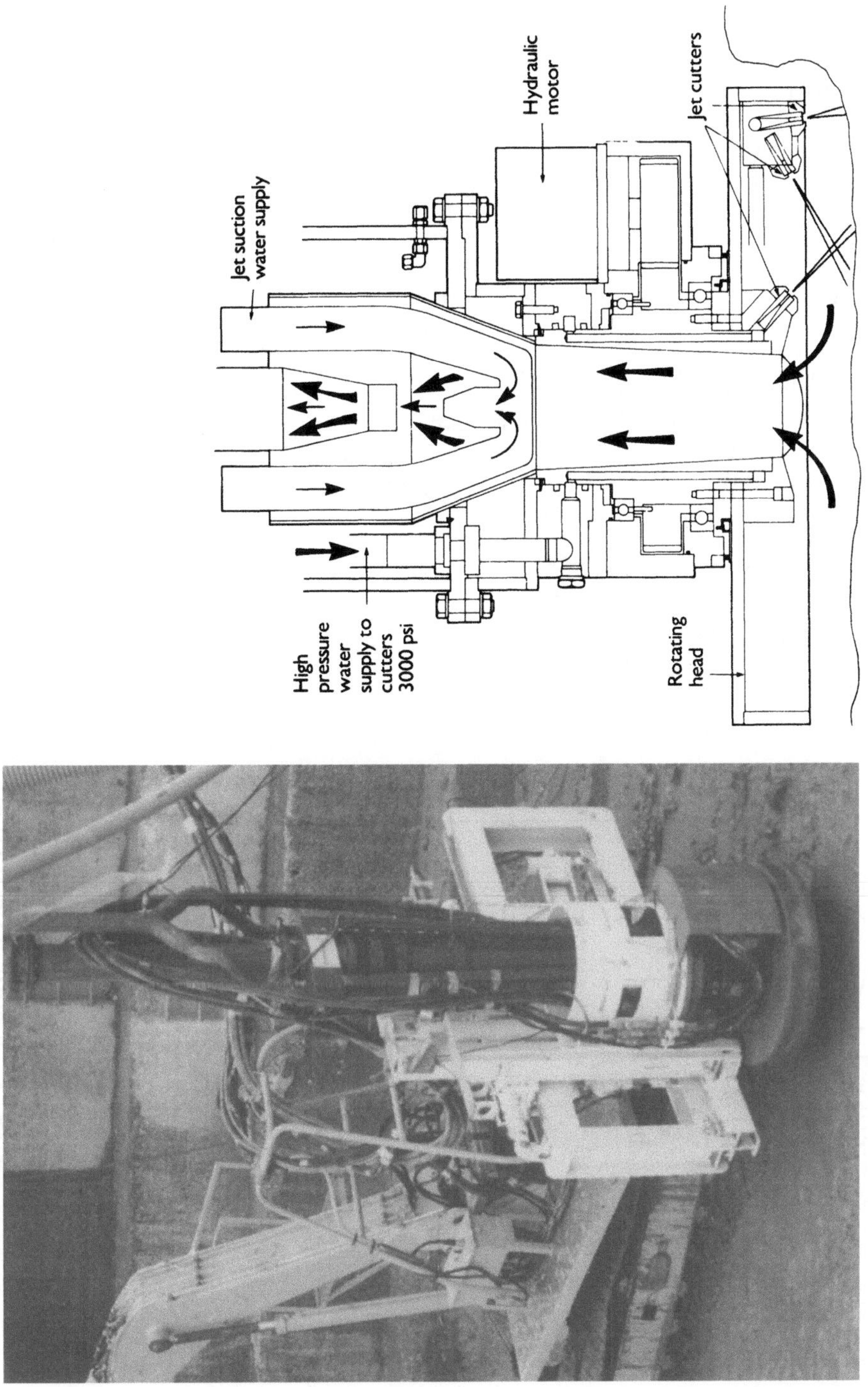

Fig. 6 Schematic of clay cutter and Blyth trial using a subsea crawler

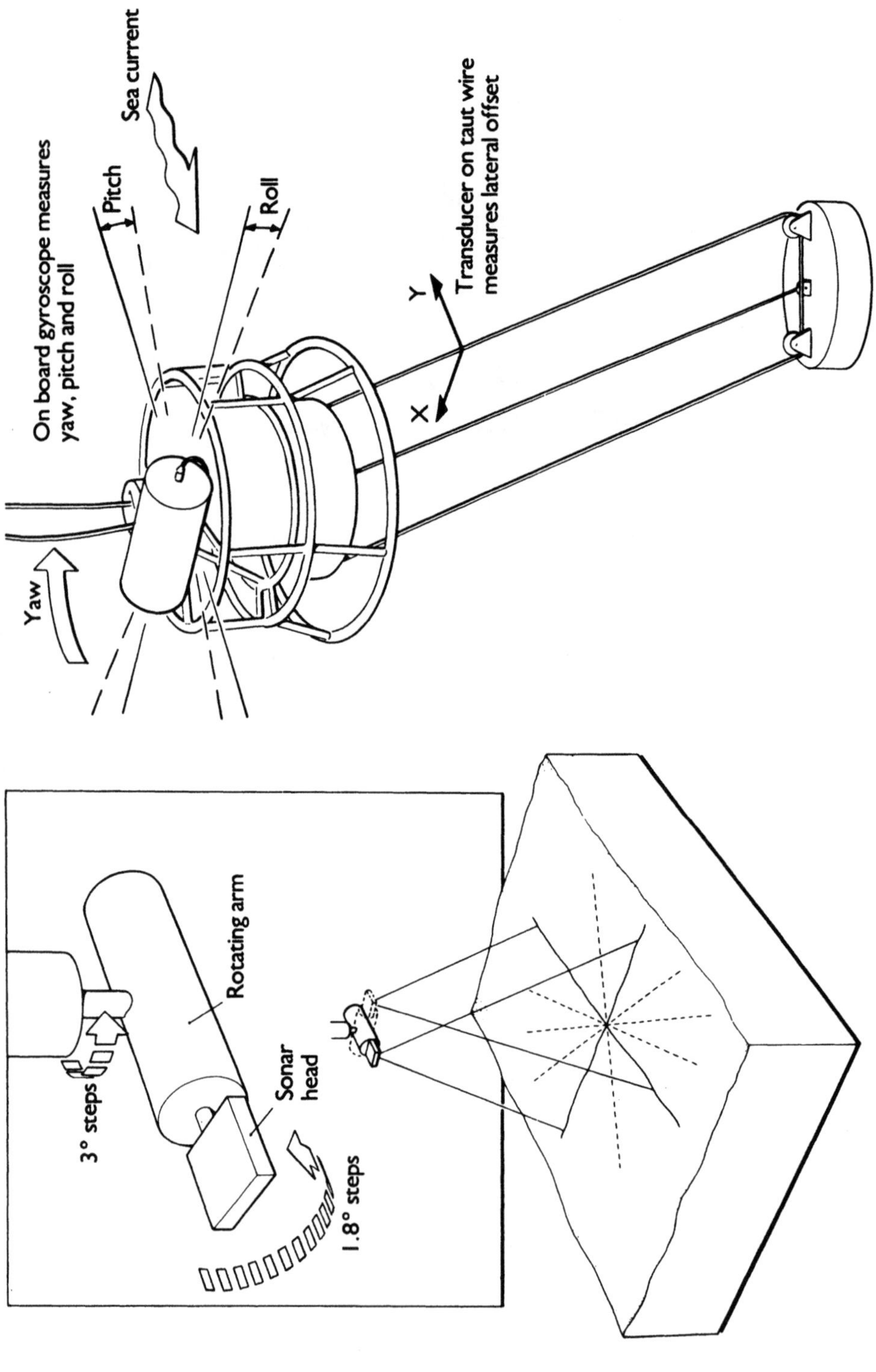

Fig. 7 Scanning pattern and buoy compensation for sonar graphics

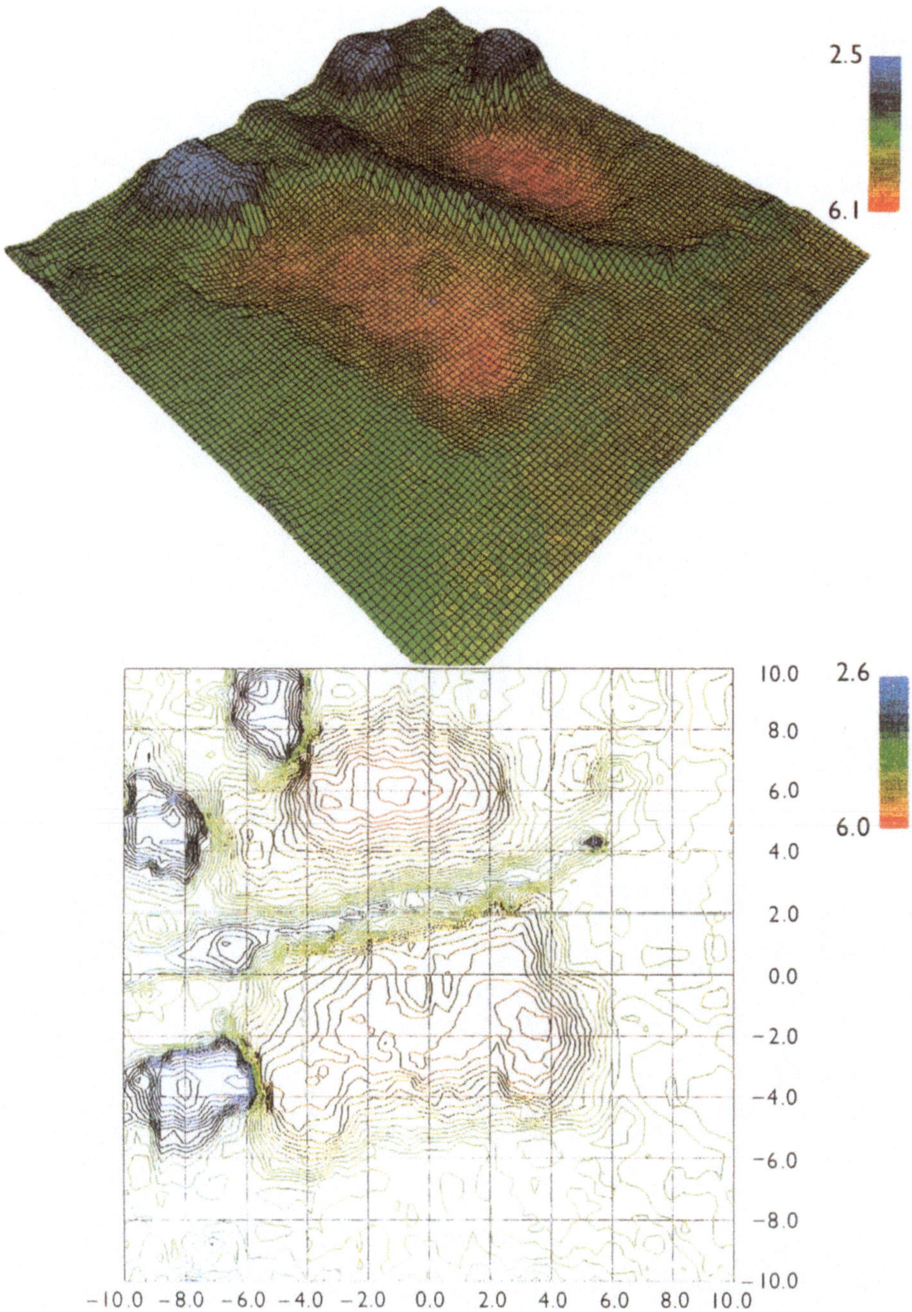

Fig. 8 3D sonar graphics plot and contour map of excavated 36 inch submarine pipeline

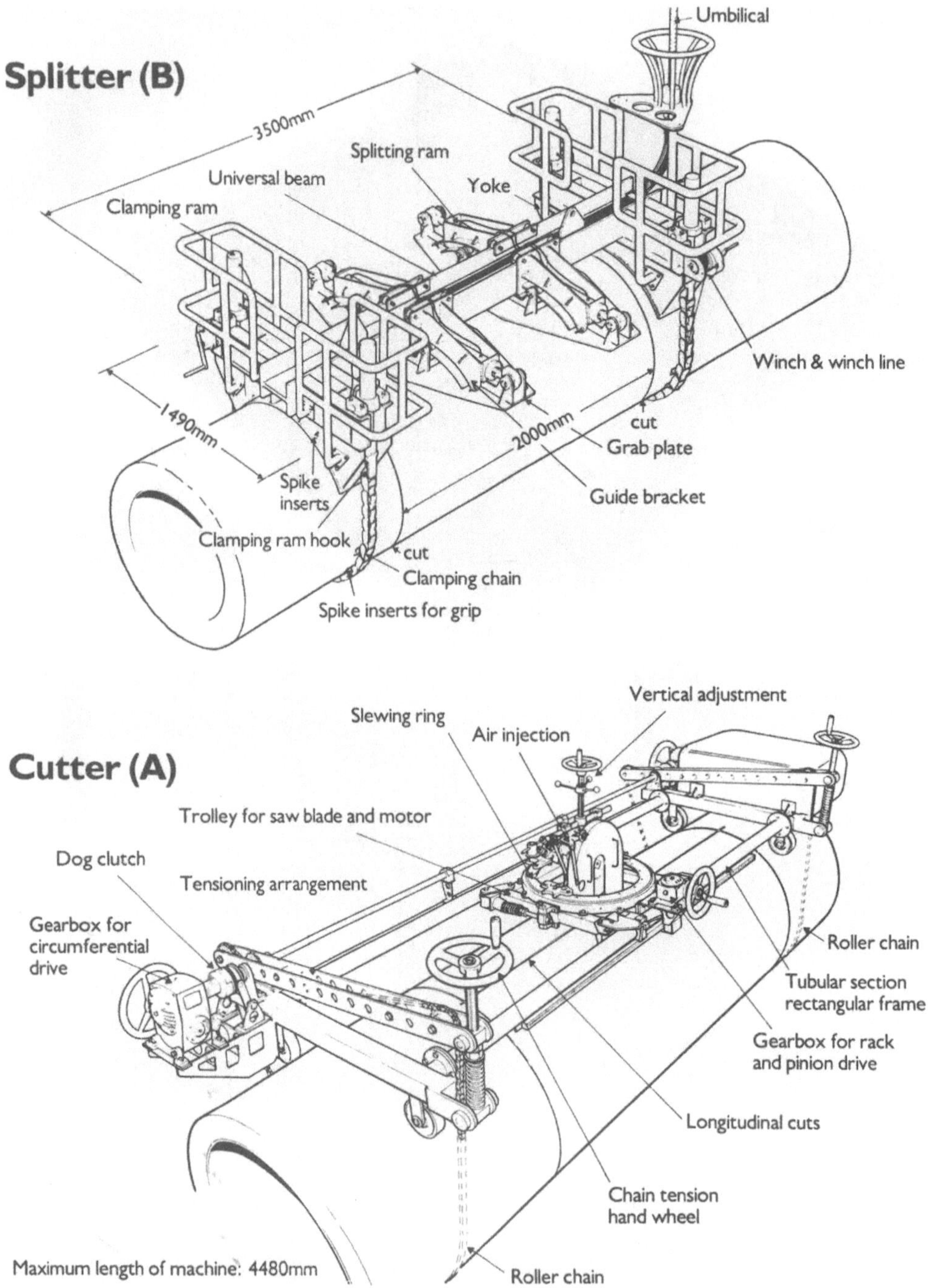

Fig. 9 Concrete removal system

Fig. 10 Friction / pig trials at Blyth

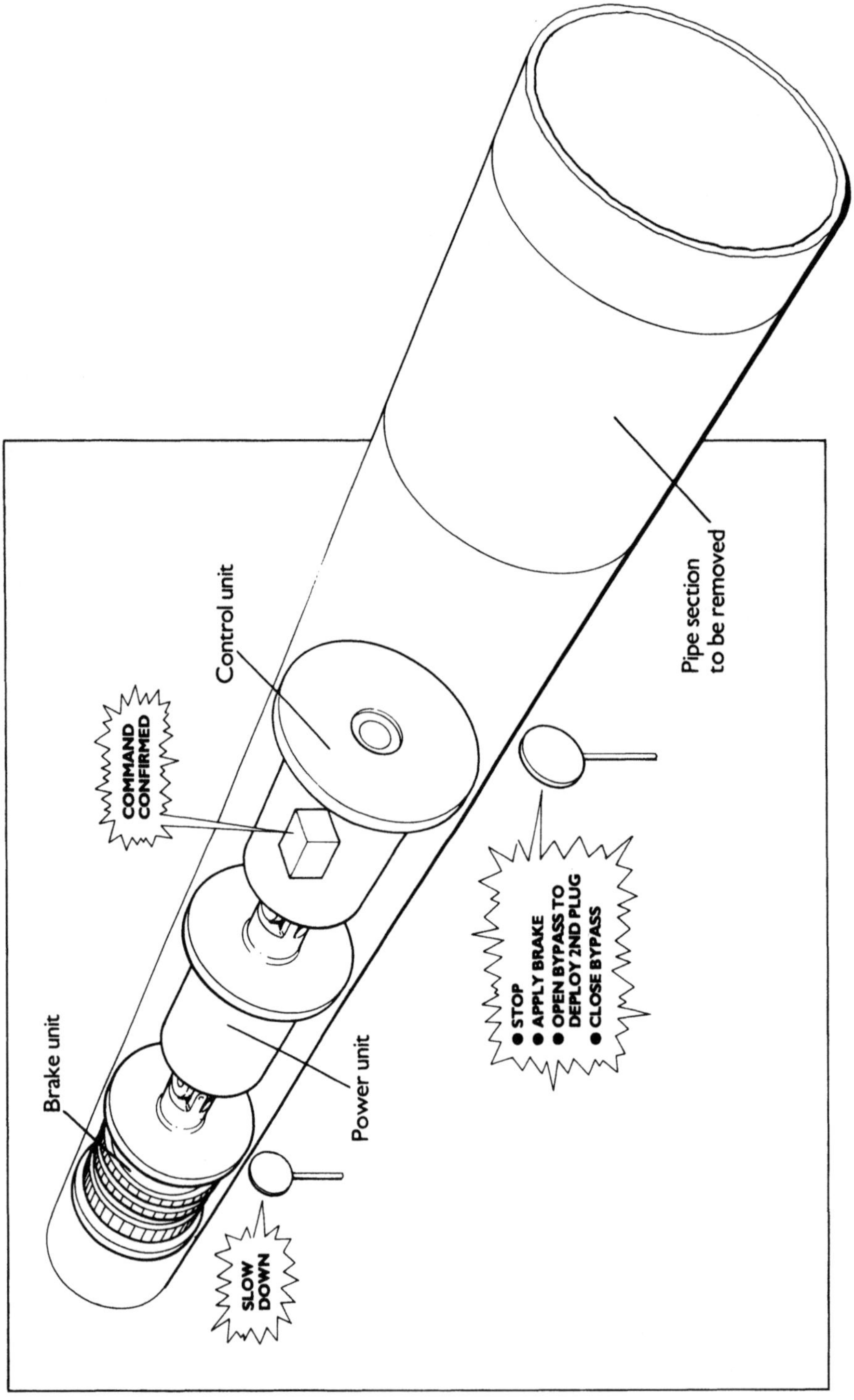

Fig. 11 Schematic of through wall communication for barrier pigs

Splitting the concrete panels away from the pipe is achieved by four hydraulically operated 'grab plates' located in a previously cut slot at the 12 o'clock position. Using a four way split power-pack ensured that all four rams moved at the same rate regardless of the pressure in each ram which prevents the panels jamming as they are prised off the pipe.

Results from the trials show that using the British Gas technique 2m concrete panels can be removed from the pipe in 4 to 6 hours depending upon visibility.

5.0 PIPELINE ISOLATION

The aim of deploying flow stopping devices is to isolate a section of pipeline and prevent seawater from flooding the pipeline when the damaged section is removed, otherwise the pipeline would subsequently require extensive de-watering and drying. A prime requirement for flow stopping systems is to resist the differential pressure between the seawater and the gas in the pipeline and create a temporary leak-tight barrier.

At present few systems have been fully proven in service for this duty and the offshore industry is still concentrating a considerable amount of effort on the problem. Outlined below are various techniques that are presently being evaluated and developed.

HOT TAP STOPPLE

British Gas in conjunction with other field operators have funded a successful trial into hot tapping a fitting onto the pipeline under hyperbaric conditions. The fitting was designed to accept a modified 36 inch folding stopple with a diver-inserted pig to give a double block and bleed facility. This was successfully demonstrated underwater at the Blyth Facility and a variation of this technology has recently been used offshore.

FRICTION PIGGING

Normally friction pigs are simply a series of oversized polyurethane discs which resist differential pressure by friction forces from the pipe wall, Figure 10. The pigs are forced into the pipeline using a hydraulic fluid, being usually launched from the platform. However they can only be deployed a few hundred metres before wear reduces their effectiveness.

Experiments with soft discs which are 4% oversize relative to the pipe's nominal bore produces a 2.5 bar differential pressure (DP) and multiple pigs can be deployed to provide a higher DP capability. Other operators have deployed friction pigs in nominally constant bore pipelines. British Gas however, were faced with significant changes in the pipeline bore dimensions which justified the construction of a full scale test section. The test programme provided the actual drive pressures for different disc combinations as the pig traversed though a simulated riser, valves, bends and variable pipeline bores. Appendix 3 gives details of the pig drive pressure for a given disc combination as it traverses the test section.

This variant of friction pigging proved ideal to isolate the pipeline for repair and installation work on the pipeline near the platform (or shoreline) or for a riser repair, but it is obviously unsuitable for maintenance work performed many kilometres away.

PIGGABLE BARRIERS

The ideal solution is to deploy a remotely controlled pig that can travel the length of the pipeline and can be positioned at any point along it.

At present a number of companies have developed pigging systems that contain sealing and braking facilities, but to date they have only been reliably deployed using a tethered umbilical which again restricts the distance to around 300m from the launch point. The Shell Spherical pig[5] is the exception to the above statement, but it was primarily designed for pipelines with 1D bends and its DP was restricted to 3.0 bar per sphere at 20 inches diameter.

Various methods of remote communication with pigs are available, eg acoustic, nuclear, and electromagnetic. The main problem is however to get good quality signals after passing through water, seabed, reinforced concrete, the pipewall and a gas interface, Figure 11.

To overcome these difficulties British Gas are working on magnetic systems with trials commencing later this year.

The aim is to establish remote communication with a moving pig using signals from a control ship via transponder type devices placed near the pipeline. As the pig passes each device, commands can be transmitted and received from the pig. Once stationary, the communication package could inflate the seals and monitor the differential pressure. Similar technology could also have a role in the onshore gas supply system.

6.0 REPAIR ANALYSIS

A best estimate for repairing a 36 inch diameter buried submarine pipeline is 66 days, with the details given in Appendix 2. This takes into account every step from locating and mobilising a surface support vessel to completing the hyperbaric welds. Also taken into account is the lost time due to poor weather and reduced productivity due to working in a 2 knot seabed current.

Using innovative technology this may be reduced to 28 days. These times will be further influenced by the length of pipeline that has to be repaired. Damage over a length exceeding 10 diameters will take considerably longer.

Times to commission a flooded pipeline have not been included because these can vary enormously depending upon various pipeline parameters. A time of 20 days should be budgeted for, especially if vacuum drying is used. If flow stopping techniques are used then this would drop to around 3 to 5 days.

CONCLUSION

Analysis of the overall operation of repairing a buried submarine pipeline that requires the defective length to be cut out, can take up to 66 days using existing repair methods. This time can be substantially reduced to about 28 days by the deployment of innovative techniques in the areas of seabed excavation, surveying and concrete weight coating removal.

Further reductions are possible if flow stopping techniques are used rather than allowing sea water to enter the pipeline.

ACKNOWLEDGEMENTS

The authors wish to thank British Gas for permission to publish this paper and to the significant contribution from the Facilities Engineering Department of British Gas Exploration and Production.

The authors also wish to thank the Institution of Gas Engineers for their kind permission to reproduce this paper which was originally given at the 1992 Autumn Conference.

LITERATURE REFERENCES

REFERENCE 1 Cotton, H. Protevin Lecture, Welding Underwater and in the Splash Zone - A Review. Underwater Welding, Proceedings of International Conference Held Trondheim, Norway, 27-28 June 1983.

REFERENCE 2 Nixon, J.H. Underwater Welding - A Review. A.T.M.A. Conference, Paris, December 1982.

REFERENCE 3 Thomas, R.C. and Wright, N. The British Gas Subsea Engineering Centre - the Role of Underwater Testing to Assist in the Evaluation, Development and Selection of the Facilities Engineering Department of British Gas Exploration and Production.

REFERENCE 4 Wright, N. Gibson, I. The British Gas Concrete Weight Coating Removal System for Subsea Pipelines. Presented to the Offshore Inspection, Repair and Maintenance Conference, Aberdeen, November 1988.

REFERENCE 5 Newman, W J. Kontou, T. Aldeen, L N. Pipeline Isolation Using Remotely Activated Spheres. Presented to South Eastern Branch Pipeline Industry Guild, 28 October 1992.

Concrete weight coating

Thickness	106mm (4.17 in) – 135mm (5.31 in)
Type of reinforcement	**(1) Circumferential** 8mm (0.31 in) dia. bar in a continuous spiral at 100mm (3.94 in) – 150mm (5.9 in) pitch. **(2) Longitudinal** 24 x 5mm (0.2) dia. bars equally spaced around circumference.
Position of reinforcement	Within the middle third of coating thickness.
Density	3040 Kg/m^3 (190 lb ft^3)
Specification	CP. 110, 1972 and BGC/PS/CW9 Pt 2 Issue 2

Steel pipeline

Outside diameter	914.4mm (36 in)
Wall thickness	15.88mm (0.625 in)
Manufacturing specification	A.P.I. 5LX with BGC BGC/PS/LX1 and GC(E) A56U/20/00/9166 Rev. C
Steel grade	X65

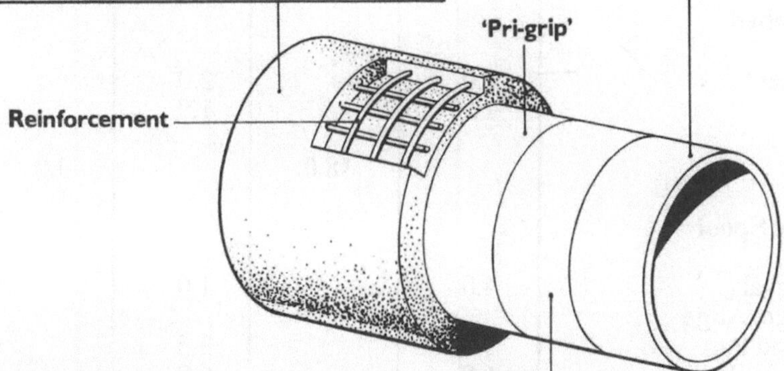

External coating

Type	Fusion bonded resin powder epoxy.
Nominal thickness	0.6mm (0.024 in) – 0.75mm(0.03 in)
Average density	1400 Kg/m^3 (97.4 lb/ft^3)
Specification	BGC/PS/CW6 and PD/EPC/1

Appendix 1 Pipe coating specification

APPENDIX 2

REPAIR TIMES FOR A 36 INCH DIAMETER BURIED PIPELINE

Activity Description for a 10 m Spool Repair in Rough Field Clay Seabed	Existing		Improved	
	Task Time	Total Time	Task Time	Total Time
	Days	Days	Days	Days
1. Mobilise Vessel, Damage Assessment of Pipe	6.0		5.0*	
Sub Total		**6.0**		**5.0**
2. Excavate Seabed				
Set Up Equipment	1.0		2.0	
Excavate	17.0		1.0	
Sub Total		**18.0**		**3.0**
3. Tie in Repair Spool				
Concrete Removal	4.0		1.0	
Cut Out Damage, Align Repair Spool and Pup	3.5		3.5	
Three Hyperbaric Welds	6.0		6.0	
Sub Total		**13.5**		**10.5**
TOTAL FOR SECTION 1, 2 & 3		**37.5**		**18.5**
4. Environmental Down-Time				
Weather (25% of total)	9.5		4.5	
Sea Currents (50% of total)	19.0		5.0**	
Sub Total Average values are used for weather down-time since they will vary from season to season,		**28.5**		**9.5**
Overall Total for Repair		**66.0**		**28.0**

*Reduced due to use of Jet Prop Excavator to Uncover Pipe.
**Reduced due to use of Cofferdam.

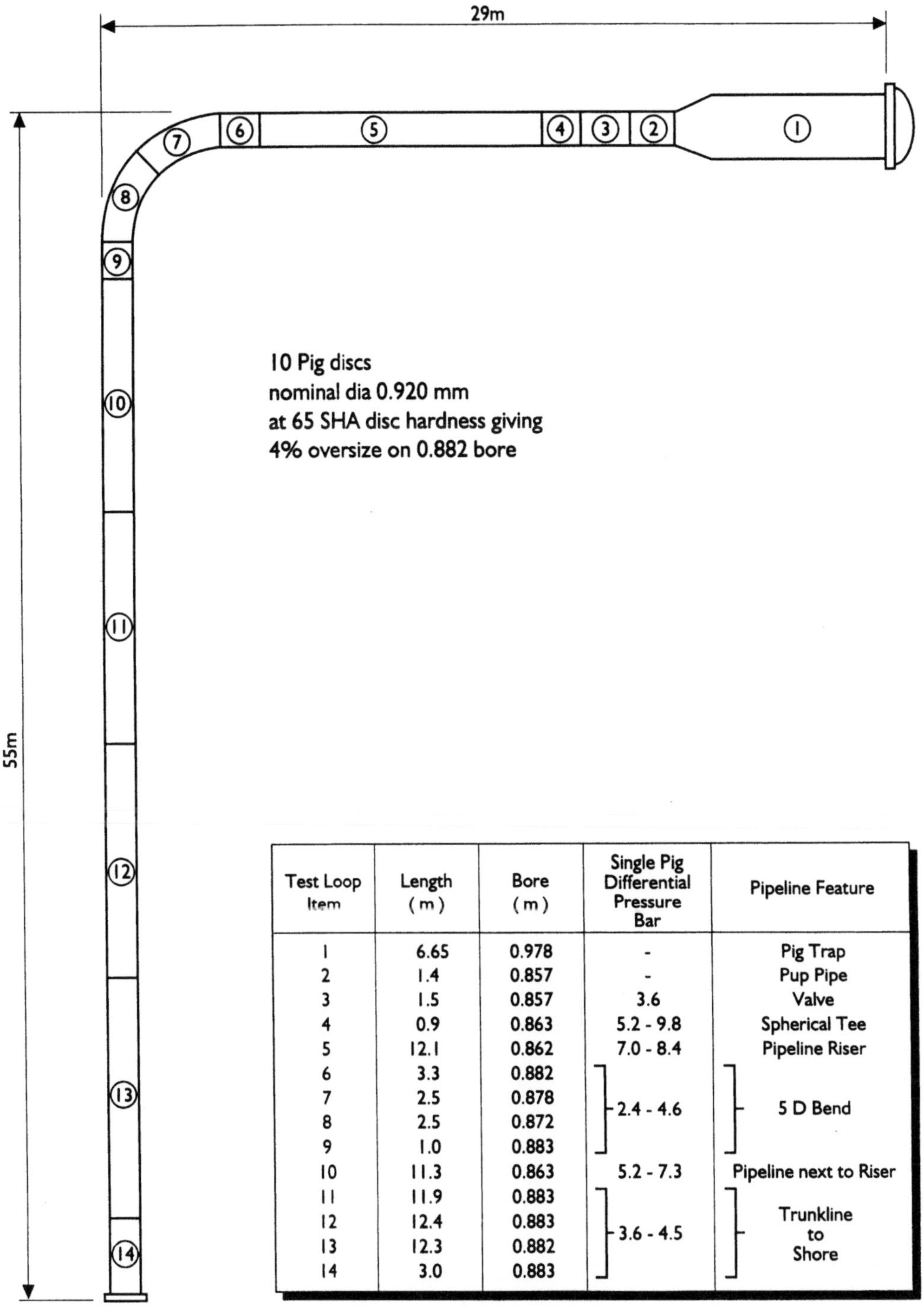

Test Loop Item	Length (m)	Bore (m)	Single Pig Differential Pressure Bar	Pipeline Feature
1	6.65	0.978	-	Pig Trap
2	1.4	0.857	-	Pup Pipe
3	1.5	0.857	3.6	Valve
4	0.9	0.863	5.2 - 9.8	Spherical Tee
5	12.1	0.862	7.0 - 8.4	Pipeline Riser
6	3.3	0.882	2.4 - 4.6	5 D Bend
7	2.5	0.878		
8	2.5	0.872		
9	1.0	0.883		
10	11.3	0.863	5.2 - 7.3	Pipeline next to Riser
11	11.9	0.883	3.6 - 4.5	Trunkline to Shore
12	12.4	0.883		
13	12.3	0.882		
14	3.0	0.883		

Appendix 3 Friction pig test data

MIX
Papier aus verantwortungsvollen Quellen
Paper from responsible sources
FSC® C105338

If you have any concerns about our products,
you can contact us on
ProductSafety@springernature.com

In case Publisher is established outside the EU,
the EU authorized representative is:
Springer Nature Customer Service Center GmbH
Europaplatz 3, 69115 Heidelberg, Germany

Printed by Libri Plureos GmbH
in Hamburg, Germany